NAMING ORGANIC COMPOUNDS
A Systematic Instruction Manual

NAMING ORGANIC
COMPOUNDS
A Systematic
Instruction Manual
2nd edition

E. W. GODLY, B.Sc., C.Chem., MRSC
Formerly Head of the Nomenclature Subdivision of
Laboratory of the Government Chemist
Teddington, Middlesex

ELLIS HORWOOD

for

and

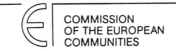

COMMISSION
OF THE EUROPEAN
COMMUNITIES

First published 1989
2nd edition published 1995 by
Ellis Horwood Limited
Campus 400, Maylands Avenue
Hemel Hempstead
Hertfordshire, HP2 7EZ
A division of
Simon & Schuster International Group

© Commission of the European Communities, Laboratory of the
  Government Chemist, and Ellis Horwood, 1989, 1995

Printed and bound in Great Britain by Bookcraft (Bath) Ltd.

Library of Congress Cataloging in Publication Data

Godly, E.W. (Edward W.)
  Naming organic compounds: a systematic instruction manual/E.W.
Godly.– 2nd ed.
  p.   cm. — (Ellis Horwood series in chemical information science)
  ISBN 0-13-103623-8
  1. Organic compounds – Nomenclature.   I. Laboratory of the
Government Chemist (Great Britain).   II. Commission of the European
Communities.   III. Title.   IV. Series.
QD291.G63   1995
547′.0014 – dc20   94–27486
                                                                             CIP

British Library Cataloguing-in-Publication Data

Godly, E.W.
Naming organic compounds
1. Organic compounds. Names
I. Title
547′.0014

ISBN 0-13-103623-8

1 2 3 4 5   99 98 97 96 95

Contents

Contents

Contents

Contents

Flow-diagram

Author's note

In 1981 I was asked if it would be possible to write an instruction manual on systematic organic chemical nomenclature—capable of covering 90% or so of all new compounds of interest in international trade—to be useful to readers with some knowledge of chemistry but little or none of the IUPAC Rules of Organic Nomenclature. It would not have been in any way reprehensible to have given a negative answer and only time and usage will show whether a simple 'NO' will prove ultimately to have been the correct response.

However, my feeling was that IUPAC systematic names are constructed according to rational and logically consistent principles which ought to be capable of explanation in terms comprehensible to the general reader and so I undertook the task despite the fact that no such 'stand-alone' work had to my knowledge been successfully produced.

At that time, my division of the Laboratory of the Government Chemist had been very active in producing a Euro-list of trade chemicals classified under their respective headings of the Common Customs Tariff. This list, now incorporated into the 'European Customs Inventory of Chemicals', EEC, Luxembourg, 1988, contained over 16 000 entries and its compilation had compelled us to supplement the IUPAC rules with in-house sub-rules and rationales aimed at consistent treatment of analogues. Armed with this body of experience, I felt that the task should not prove too onerous. That proved to be over-optimistic. However, trials on an earlier attempt gave a 70% success-rate and this 1988 revision has plugged a number of gaps, removed errors and inconsistencies and remedied inadequacies.

Mercifully, an instruction manual leaves no room for argument and non-preferred methods of naming, however valid, are omitted from consideration. Names produced by following these instructions should prove consistent with those of the Euro-Customs Inventory, Nos. 10001–26591.

For best results, readers should beware of using their own knowledge of the subject to modify specific instructions. These should be read carefully and applied faithfully.

Not every conceivable molecular structure is covered by the book; some rarely encountered species have been omitted, for to aim close to 100% success would require a very much thicker volume even than this. However, it is hoped that it will prove effective in most cases.

If this is found not to be so, or if a reader experiences any difficulty in obtaining such a result, enquiries should be directed to the Chemical Nomenclature Advisory Service, at the Laboratory of the Government Chemist, Teddington.

I am indebted to my colleague Mr Ivor Cohen for valuable advice and help in the preparation of this revised version.

E.W.G.

1 Introduction

This work was undertaken in the context of the application of Council Directive 79/831/EEC; of 18th September 1979, amending for the sixth time Directive 67/548/EEC on the approximation of the laws, regulations and administrative provisions relating to the classification, packaging and labelling of dangerous substances.

1.1 AIMS AND FUNCTIONS

This book is in no way intended to replace the IUPAC Rules of Organic Chemical Nomenclature, to which serious students of the subject are directed,† but rather to provide a programme or recipe for devising an acceptable name for a new chemical under those Rules, in the form of a set of procedures not requiring the skills of a nomenclature-specialist. Some knowledge of basic chemistry is, however, assumed on the part of the reader. The names so generated will, it is hoped, provide unambiguous names in an internationally acceptable style, namely that of the IUPAC Rules.

These rules have been the basis of our naming rationales because the above-cited directive requires notification of new chemical substances before they are placed on the Community market, defined in these notifications by an IUPAC name. This approach has been adopted because the IUPAC philosophy for naming chemicals offers the following practical advantages for normal communication in international trade-contexts over more highly systematized methods:

(1) they incorporate a body of useful, established trivial names, which must continue to be the obvious first-choice names in all manner of practical industrial as well as legislative and regulatory contexts,

(2) their rules are translated into several languages and are continuing to expand their international acceptance,

† *Nomenclature of Organic Chemistry*, Sections A–F & H, 1979, Pergamon.

(3) their usages and conventions tend more often than not to result in names which are comparatively neat, succinct and readily intelligible to the chemical community,

(4) the IUPAC Rules have remained very largely the same for the last 30 years.

Against all that it must be admitted that, because of their traditional function of codifying various 'respectable' current nomenclature-practices, IUPAC Rules frequently allow more than one name for a given chemical. Accordingly, with a view to harmonizing the designation of new chemicals for the application of Directive 79/831/EEC, the methodology leading to a single, preferred IUPAC name had to be developed. To this end, *ad hoc* sub-rules devised for the purpose of obtaining consistency in the selection of preferred names for trade-chemicals in '*Classification of Chemicals in the Customs Tariff of the European Communities*'† have been used—with some amplification—in this book also.

In view of limitations on size and convenient handling, such a treatment could not hope to cover all possible structures, including those not yet conceived. Accordingly deliberate omissions of rarely seen structure-elements have been made and no treatment has been attempted for sections of the rules which are particularly complicated. It is, however, hoped that 90–95% of naming problems likely to be encountered will be found to be covered adequately by following the detailed instructions provided. Questions on naming arising either from the 5–10% area not covered, or for any other reason should be directed to the

Chemical Nomenclature Advisory Service (CNAS),
Laboratory of the Government Chemist,
Department of Trade and Industry,
Queens Road, Teddington,
Middlesex, TW11 0LY, UK

1.2   HOW TO USE THE BOOK

It is advisable to read Appendix D before attempting to name any but the simplest structures. Having done that, proceed as follows:

Unfold the **flow-diagram** at the back of the book and, commencing at the starting-point on the left, follow the track by addressing each question to the known structure of the single chemical under consideration, following the path according to the answer at each stage. To do this effectively, it may be necessary to rotate the structural formula about any of the three Cartesian axes for purposes of comparison with the structures printed in the book. Moreover, there are a number of ways to depict the same molecular situation and non-chemists may fail to appreciate that they are, in fact, identical. As an example, the following methods of depicting 3-methyl-4-propylhept-3-ene by no means exhaust all the possibilities:

(a)   (b)   (c)

$$CH_3CH_2CH_2-C(=C(\overset{\overset{\displaystyle CH_3}{|}}{\underset{\underset{\displaystyle CH_2CH_2CH_3}{|}}{}})-CH_2CH_3$$

(b) $Pr_2C=C$ with Pr, Me above and Pr, Et below

(c) $Pr_2C=C(Me)(Et)$

† Ed. by Forcheri, S. and de Rijk, J. R., EEC, Brussels, 1981.

(d)

$(CH_3.CH_2.CH_2.)_2C:C(CH_3).CH_2.CH_3$

(e)

As a second example, here are some ways of showing the aspirin molecule:

(f)

(g)

(h)

$C_6H_5(o\text{-}OAc)\text{---}CO_2H$

(i)

(j)

In fused-ring systems alignments can vary and the five-membered ring may appear as a regular pentagon, a 'house' or an inverted 'house':

(k)

(l)

(m)

(n)

Any rotation of a molecular structure as a whole is permitted for purposes of comparison with the reference-structures, remembering that formally there is in general free-rotation about all single bonds but none about multiple bonds. Reflection of the structure in a mirror (wherever placed relative to the structure) is not a permitted rearrangement, however.

If such potential variations in presentation are understood, it should be a straightforward procedure to follow the appropriate paths in the **flow-diagram**. Eventually the searcher will, having answered the relevant questions, arrive at a terminal panel, heavily framed and hearing an arabic numeral corresponding to the section number that carries specific instructions for constructing the name. This is a mechanical exercise requiring no special expertise in chemical nomenclature. However, it is appropriate to explain the basis of name selection in this scheme and this is done in section 1.3.

On reaching the appropriate numbered box at the branch-end of the **flow-diagram**, the reader should turn to the section number concerned in the Table of Contents and select the sub-section appropriate to the molecular structure so as to avoid unnecessary searching. Each such sub-section carries detailed instructions for building the name but sometimes the reader is asked to refer to other pages for lists of relevant trivial names or for general instructions such as those for numbering. There may even be redirection to another box-number.

In complicated cases it will be advisable to read also Appendix C (in particular C-4). Any stereo-symbols should be added, as appropriate, after reading section 2. This operation may well necessitate the addition of further enclosing marks.

Finally, it is a useful procedure to see if the name derived by using this book yields the correct structure unambiguously. This can be checked by a colleague or, after a suitable time-lapse, by the reader.

1.3 SELECTION PRINCIPLES

Systematic chemical names are based on the concept of a parent structure, be it a chain, a ring-system (of one or more rings), or even a ring-system with attached chain or chains (see section 8). The choice is made according to hierarchical rules and these form the subject of section 1 and the Yellow Pages, which should be consulted before going on to apply the rest of the book to a particular structure. Use is made of appropriate trivial names and the trivial name covering the largest parent structure is generally preferred, e.g. 1,3-xylene rather than 1,3-dimethylbenzene. At the same time, preference is accorded to functional groups such as acid, ketone, nitrile, etc., again arranged in a hierarchical order, so that the preferred name results from an interplay of these determining principles, and it is this which underlies the **flow diagram's** construction. Once the parent-name, together with its principal functional group or groups has been decided, the rest of the structural elements are built in to the name by means of appropriate infixes or prefixes so that the finished name conveys the entire chemical structure unambiguously–and nothing else, i.e. with no uncertainties or false indications. Like entities are, so far as is possible, collected and treated identically. Thus, we prefer α,2,3-trichlorotoluene to 2,3-dichlorobenzyl chloride.

Use of trivial names

In addition to the advantages of familiarity and convenience in name-construction which certain trivial names offer, there is an additional circumstance which exploits their use. In cases of tautomerism or other molecular rearrangements, the use of a structure-based systematic name could well be over-restrictive. Thus, for example, it seems better to say 1-chlorobarbituric acid than 1-chloropyrimidine-2,4,6($1H$,$3H$,$5H$)-trione in view of the contribution from hydroxy-forms with ring-unsaturation. On the other hand, in the case of its 1,3,5,5-tetra-alkyl derivatives, in which the structure is 'frozen', the systematic name is more precise.

Substitutive nomenclature has generally been preferred in this book except where other styles, e.g. radicofunctional, are unavoidable (another reason for preferring α,2,3-trichlorotoluene in the example above). Where the IUPAC Rules have failed to cover a particular point, rationales have been developed to ensure internal consistency and these have been incorporated into the naming scheme and its associated instructions.

1.4   EDITORIAL PRACTICES AND CONVENTIONS

These will largely be explained in context, but some generalizations can be given here:

(a) Chemical names appearing in the body of a text are treated as common nouns; they do not begin with a capital letter except where specially provided for below. In indexes or catalogues initial capital letters may be used if desired. After several symbols of various kinds which may appear in front of a name it can be helpful for directing the eye to the starting letter for such purposes.

(b) Locants for atom positions on a chain or in a ring-system are conveyed by means of arabic numerals (1,2,3...) separated by commas and the set followed by a hyphen if they come at the start and flanked by hyphens if they come in the middle of a name. Thus CH_3—CBr_3 is designated as 1,1,1,-tribromoethane and CH_2Br-CCl_3 is named 2-bromo-1,1,1-trichloroethane.

Attachment of groups to atoms other than carbon can be designated by numbers or sometimes by the capital italic letter denoting the atomic symbol for that element, e.g. 1,1-dimethylpiperidine or *N*,*N*-dimethylpiperidine. The comma serves to avoid confusion between 1,2 and 12; this does not arise for *N*,*N* but the comma is still retained there for formal reasons. Positions may also be denoted by means of Greek letters and, again, commas are used to separate them. See also '**Locant-set**' in section 1.5.

(c) Spaces are introduced into chemical names only when required by the sense. They have structural significance and are not optional.

(d) Breaking the name at the end of a line is best avoided but, if unavoidable, it should be done with observance of the following conventions:

— If the break occurs where there is a space in the name anyway, add nothing at the line-end.
— If there is a hyphen in the name at the break-point, end the line with a hyphen.
— If there is no break in the name at the line-end—and for this purpose enclosing marks are considered along with letters—the following symbol should be used: ⌒ or, if this is not available, the = sign.

(e) Enclosing marks. Three kinds are used and the nesting-order is as follows:

$$\{ \; [\; (\;) \;] \; \}$$

If higher orders are needed they are repeated as often as desired. Obviously the opening marks in a name should be matched in number and kind by the closing marks. Special circumstances governing their use will be given in context.

(f) Subscript numerals. These denote multiplication of the group to which they are attached. Thus, —$[CH_2]_4$— denotes

```
     H   H   H   H
     |   |   |   |
  —C—C—C—C—
     |   |   |   |
     H   H   H   H
```

(g) Superscript numerals. These are used as distinguishing marks only. Thus, $R^1R^2R^3R^4Si$ denotes a silicon atom with four different groups attached to it. More specialized uses will be dealt with in context.

(h) Multiplying affixes. Mono is hardly ever used but di, tri, tetra, penta, etc., and bis, tris, tetrakis, etc., are used in names according to the particular situation. There is no simple rule governing which kind is used and instructions will, in each case, be explicit.

(i) Elision of vowels. Again there is no simple rule and instructions will be given in each case so that the resulting names follow the practices of the list of names for EC Trade-chemicals.†

(j) Numbering systems. There are two kinds of numbering system: systematic and trivial. The instructions will deal with both kinds as they arise but it should be noted that when the locants occur inside enclosing marks they apply only to the numbering of the system named inside those enclosing marks. Thus, in the case of 2-(2-chlorophenyl)propan-1-ol, the first '2' applies to the propanol chain, whereas the '2' inside the parentheses applies to the numbering around the attached benzene ring. Such use of parentheses obviates the need for using primes to distinguish the two numbering sequences (e.g. 2'), although this device is used in certain circumstances (e.g. for ring-numbering of anilides).

(k) Italics. These are used for certain structural prefixes, e.g. *sec*, *tert*, (but not iso), *cis*, *trans*, elemental symbols such as *S* and *N*, and letters representing polygon-sides in fusion, i.e. *a,b*, etc.

Italic characters are not taken into account for alphabetical order considerations until all those of the roman character set are found to be identical.

1.5 DEFINITIONS OF TERMS USED IN THE FLOW-DIAGRAM AND EXPLANATORY NOTES

Aromatic. This term defines a special kind of unsaturated ring-system. For present purposes it suffices to say that it refers to the bonding situation in a six-membered ring containing three double bonds in alternating positions. i.e.

Any ring-system containing at least one such ring is 'aromatic'.

Chain. This is the term used to describe atoms joined together in a non-cyclic manner. Chains may be straight or branched and the term is used even when the chain has only one member.

Close association. This term has been coined in respect of ring-systems for the purpose of making a pathway-choice in the **flow-diagram**. Examples are shown below:

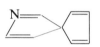

one shared atom   two shared atoms   one atom common
                                                to three rings

† See footnote to p. 2.

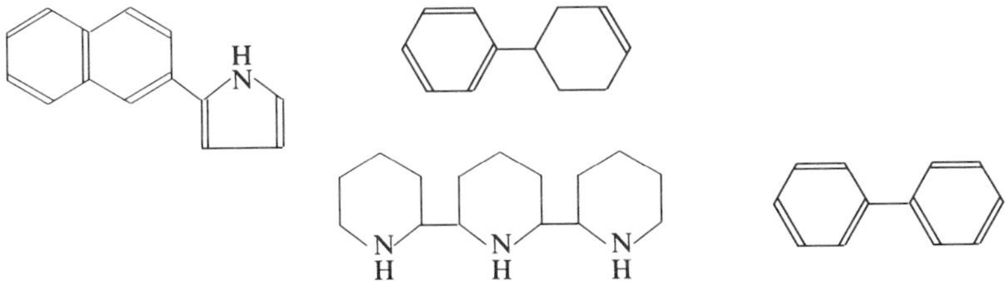

direct bonding ring-to-ring

However, the naming of such diverse structures is not always straightforward and various criteria have to be considered, as set out in the rules governing the seniority of ring-systems (see yellow pages section A).

In the following cases the rings/ring-systems are not in 'close association':

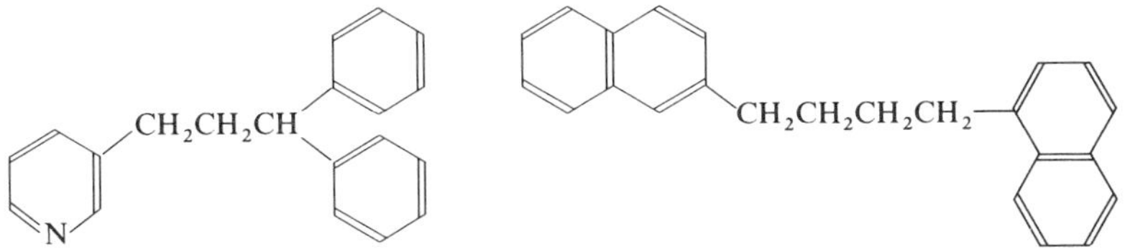

In the last example, the two fused rings are, in each case, in close association but the two resulting naphthalenes are not.

Collected. In the **flow-diagram** this means capable of being treated together in the name-ending. This can come about in various ways and the detailed procedure for each is given in its appropriate context, but briefly they are as follows:

(i) the principal group (PG) can be joined directly at several places in the same chain, as in the following example:

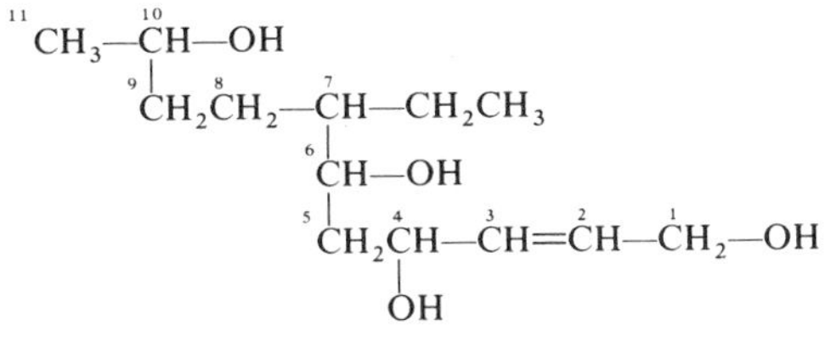

When this is the case, such groups can be collected in the name-ending by means of a multiplicative prefix (di-, tri-, etc.). In the above case the –OH groups are the PGs because they are the only kind of group present from Table 1, and they are each denoted in the name by the suffix 'ol'. They are all directly attached to the chain numbered 1–11 and so they are collected in the name: 7-ethylundec-2-ene-1,4,6,10-tetrol ('a' of tetra is elided before 'o').

However, in the case of HO—CH_2—NH—CH_2CH_2OH the –NH– group has the effect that, under the naming procedures of this book, the –OH groups are not collectable and only one can be the subject of the name-ending. (Remember that –OH is 'senior' to –NHR for the purpose of choosing the PG from Table 1). The –OH group on the two-carbon chain is preferred to that on the one-carbon chain and so the name is 2-[(hydroxymethyl)amino]ethanol.

In some cases PGs may be collected in a trivial name. For example:

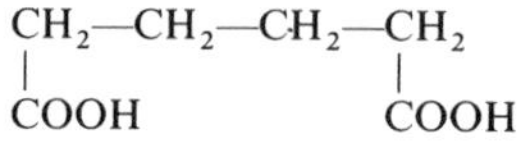

$$\begin{array}{cccc} CH_2-CH_2-CH_2-CH_2 \\ | \qquad\qquad\quad | \\ COOH \qquad\qquad COOH \end{array}$$

is named adipic acid. In this case the PG is –COOH, it occurs twice in the structure and both are directly attached to the same chain of carbon atoms. If the PG occurs more than twice, then preference is given to the name which collects the maximum possible number of PGs in the name-ending. For example, in the following case:

$$\begin{array}{c} COOH \\ | \\ CH_2-CH_2-CH-CH_2 \\ | \qquad\qquad\quad | \\ COOH \qquad\quad COOH \end{array}$$

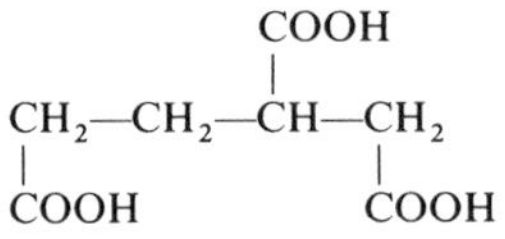

the name 3-carboxyadipic acid collects only two PGs, even though it would give the correct structure. On the other hand, the name butane-1,2,4-tricarboxylic acid collects three and is therefore preferred. (Notice how the choice of 'name-basis' affects numbering).

(ii) The PG may be directly attached to the same ring or ring-system, when similar considerations apply. For example, in the following case:

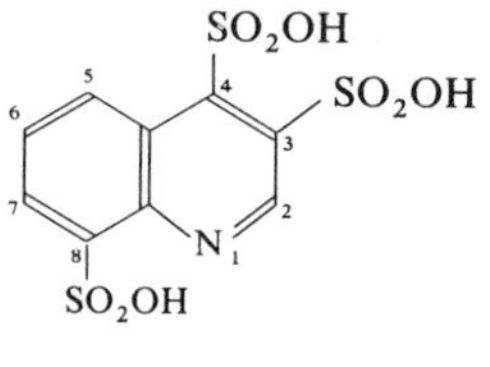

the ring-system is called quinoline and it has the fixed numbering shown. The PG is –SO$_2$OH and it is called 'sulphonic acid'. Thus, the name can collect all three PGs in the suffix:

quinoline-3,4,8-trisulphonic acid

However, in the following case:

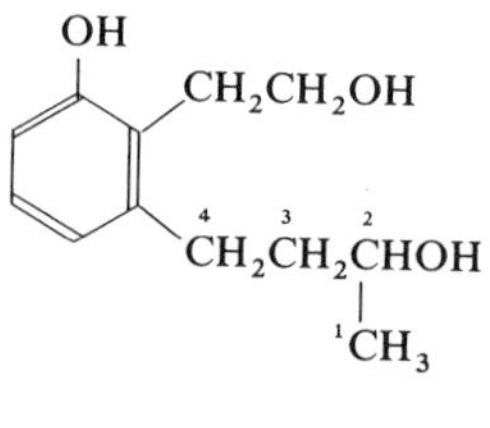

there is no way to collect the three –OH groups in a single name-ending. Accordingly, the name is based on the OH-group attached to the longer of the two side-chains, viz.

4-[3-hydroxy-2-(2-hydroxyethyl)phenyl]butan-2-ol
(alcohol preferred to phenol in Table 1 and 4C-chain preferred to 2C-chain).

(iii) The PG(s) may occur on a chain or a ring-system. This constitutes a combination known as the parent functional structure (PFS), q.v. This itself may be joined two or more times to a symmetrical central group or ring-system which is di- or polyvalent.

The following example shows two ethanol molecules joined directly to a central benzene ring. As the chain (in this case) carrying the PG is repeated and the central structure can be expressed as a divalent radical, the two –OH groups can be collected in the name-ending:

2,2'-(o-phenylene)diethanol.

This name signifies that ethanol is joined at its number-2 carbon atom in each case to a central divalent radical called o-phenylene (the structure on the right).

$$\overset{2}{C}H_2\overset{1}{C}H_2OH$$
$$\overset{2'}{C}H_2\overset{1'}{C}H_2OH$$

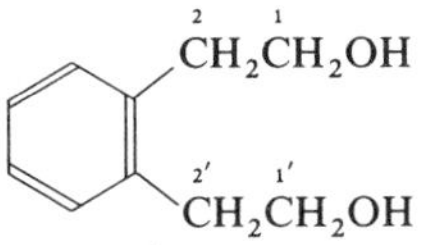

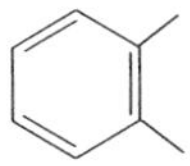

Moreover, $HO—CH_2CH_2—NH—CH_2CH_2OH$, unlike the second example in part (i), has—owing to its symmetry—two –OH groups which are collectable in the name 2,2'-iminodiethanol.

These assembly-names are discussed more fully in context but the conditions for collection in the name-ending are that there should be a di- or polyvalent radical capable of conveying the central group or structure to which the PG-bearing molecular residues are directly attached, and that these residues are the same when unsubstituted.

Thus, in contrast to the last example, the extra oxygen atom in the next case has the effect that the –OH groups are no longer collectable, because there is no divalent radical available to name the structure on the right.

$$OCH_2CH_2OH$$
$$CH_2CH_2OH$$

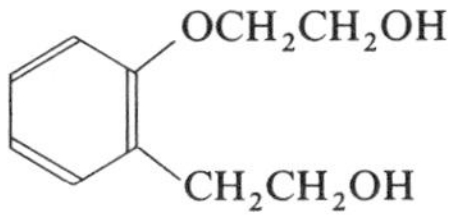

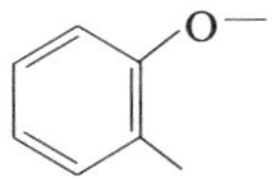

Cumulative double bonds. These are a contiguous arrangement, as shown:

$$>C=C=C=C=C=C<$$

Functional group (FG). For the purposes of the **flow-diagram**, this means any of the heteroatomic groups (with or without carbon atoms as the case may be) listed in Table 1. When two or more FGs are contiguous a new FG is usually thereby generated, e.g. $>C=O$ joined to –OH is –COOH.

Fused rings are those rings having two adjacent atoms shared between pairs of neighbouring rings. Atoms common to two rings are called 'angular' or 'bridgehead' atoms, and those common to three rings are called 'interior' atoms; they need not necessarily be carbon, e.g.

Fused   Bridged

For the purposes of Box (xxiv) of the **flow-diagram**, sharing between two rings of more than two atoms at a time is regarded not as fusion but as bridging (see section 19).

Heteroatoms. For the purposes of the book, these are atoms other than carbon and hydrogen.

Locant. The locant is the numeral, Greek letter or italic elemental symbol which denotes the position on a parent structure of an attached group (atomic or molecular). In a name the locant, followed by a hyphen, precedes the name of its attached group.

Locant-set. When the same group is multiply attached directly to a parent structure, the locants for each such position of attachment are cited in the order:Elemental symbol(s), Greek letter(s) in alphabetical order, and numerals in ascending order—all separated by commas. The set is followed by a hyphen and then the appropriate multiplicative prefix.

Example

$N,\alpha',2,2',3',5'$-Hexabromoacet-*p*-toluidide

Mancunide structures. Ring systems containing the <u>m</u>aximum <u>n</u>umber of non-<u>c</u>umulative <u>d</u>ouble <u>b</u>onds. This term was coined from the letters underlined, the 'ni' being added for euphony. However, the term 'mancude' is also used.

Multiplicative prefix. This is a prefix signifying the number of times a group or structurally modifying process is repeated, e.g. di for 2, tri for 3, tetra for 4, penta for 5, and so on. Where these would cause ambiguity or where a repeated group is substituted, bis, tris, tetrakis, etc ..., have the same function. For directly linked assemblies of identical ring-systems bi, ter, quater, etc ..., are used.

ncdb. Non-cumulative double bonds, i.e. arrangements of double bonds which have at least one single bond between them.

Parent Functional Structure (PFS). This term applies to a unit of structure (ring-system, chain or a trivially named combination of both) together with the PG or PGs it bears – whether as part of the structural skeleton or else directly attached to it. Thus, in 1.5 (iii) p. 9

the PG in the first example is -OH ("-ol") and the PFS is CH_3CH_2OH ("ethanol").

If the structure has more than one PG in a non-collectable manner, the PFS is that having priority after application of the criteria of the Yellow Pages (these take into account the exceptional cases of a PFS whose name conveys also the presence of certain attached groups).

The PFS provides the basis for the names of substituted derivatives and it must accordingly have accepted defined numbering.

Examples

(1)

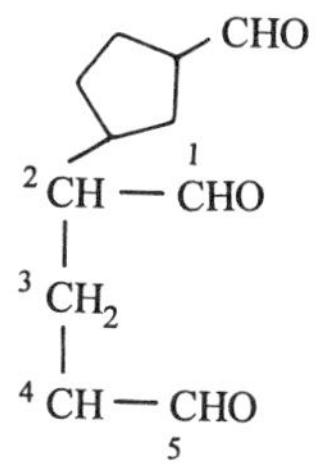

Here the PG is –CHO ('aldehyde'). The PFS is glutaraldehyde and the full name 2-(3-formylcyclopentyl)glutaraldehyde.

(2)

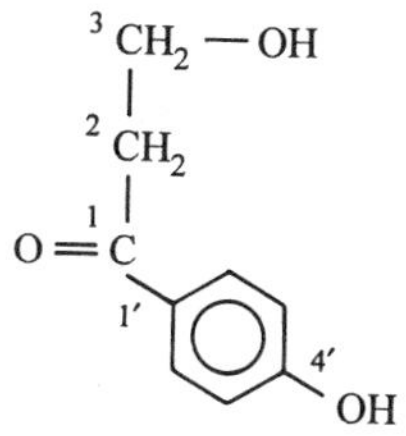

Here the PG is $C=O$ ("-one") and the PFS is propiophenone. Full name: 3,4′-dihydroxypropiophenone.

Principal group (PG). When two or more different groups from Table 1 are present in a chemical structure, that group which is nearest to the top of the list is the 'principal group'. It may be found to be present more than once in a given structure, in which case it will still be decisive in forming the name, but only after further **pathway-exploration** in the **flow-diagram**, in which it is abbreviated to PG.

Saturated. This term describes compounds with no double or triple bonds present. If such bonds are present, the compound is unsaturated.

Side-chain. Carbon atoms of a cyclic compound which are not occupying ring-positions are referred to as side-chains, even in the limiting case of a single carbon atom. Thus, toluene consists of a benzene-ring joined to a side-chain consisting of a methyl group, and styrene consists of a benzene-ring joined to a two-carbon chain containing an ethylenic bond.

In a branched chain compound, chains attached to the principal chain (as decided by the criteria of section B, yellow pages) are called 'side-chains'.

Skeletal structure. This is the set of catenated or cyclically connected atoms forming a discrete, nameable structural entity, disregarding bond-order and attached groups.

Example

In

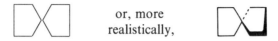

the skeletal structures are

and

Spiro-compounds. These are compounds containing a special kind of double-ring system which is characterized by the presence of a single atom shared by two rings, e.g.

or, more
realistically,

The right-hand diagram above illustrates the usual convention of depicting bonds lying in the plane of the paper by ordinary lines, those projecting towards the reader by thickened lines and those pointing away behind the plane of the paper by dotted lines. This shows the typical spatial situation at a spiro-junction.

Stereoisomers. These are, for most practical purposes, of two types:

(a) geometrical isomers, in which at least two of the groups or atoms attached to the carbon atoms at opposite ends of an ethylenic linkage are different:

$$ \underset{Y}{\overset{X}{>}} C=C \underset{Y}{\overset{X}{<}} \quad \text{or} \quad \underset{Y}{\overset{X}{>}} C=C \underset{Q}{\overset{Z}{<}} $$

often described as *cis-* or *trans-*isomers; and

(b) optical isomers, in which a single carbon atom has four different atoms or groups attached to it, giving rise to the possibility that a single isomer may be involved having a unique tetrahedral arrangement which is non-superimposable on its mirror-image.

 In deciding whether attached groups are identical or not it is necessary to explore along attached bonds, moving outwards from the suspected central asymmetric atom. For example, in the following structure:

$$ \begin{array}{c} CH_2{-}CH_2{-}CH_2{-}Br \\ | \\ H{-}{*}C{-}CH_2OH \\ | \\ CH_2{-}CH_2{-}CH_2{-}Cl \end{array} $$

the central carbon atom(*) can be said to have one hydrogen atom and three $-CH_2-$ groups attached to it, but, if the groups joined to each of those are compared, it can be seen that they are all different—even though, in two cases, exploration has to go four places along the chain to expose the difference. The central atom is therefore an asymmetric (or *chiral*) centre, giving rise to two optical isomers.

2 Stereochemistry

The two basic types of stereoisomerism are outlined towards the end of section 1.5. In one case the arrangement of connected atoms or groups about a single centre is considered; in the other, the arrangement about the atoms at opposite ends of a double bond is considered. In both cases these connected atoms or groups have to be arranged in a hierarchical order based firstly on their atomic number and then, if necessary, on their atomic mass number. Thus, in the following case:

$$
\begin{array}{c}
CH_3 \\
| \\
H-C-Cl \\
| \\
NH_2
\end{array}
$$

the central carbon atom is 'chiral' because it has four different neighbours bonded to it and their order or seniority is Cl, N, C, H.

2.1   OPTICAL ISOMERISM

The method of determining the absolute chirality for a known centre is to arrange the bonds issuing from it so that the least senior ('d' say) points away from you and the other three tetrahedrally arranged bonds are towards you. [Bonds drawn as dots or dashes are conventionally taken as lying below the plane of the paper, bold perspective wedges as projecting above it and, in such contexts, bonds drawn normally are assumed to lie in the plane of the paper. H is often cast in the role of (d). Commonly the other three are found to be C, O, N, S or Cl.] Thus, in the following case:

the direction $a \rightarrow b \rightarrow c$ (in this case Cl→N→C) is **clockwise** and so the isomer is designated (R); the alternative is designated (S).

When the same atom occurs more than once, exploration outwards along bonds must be made (as explained in the introduction) to examine their connectivity and establish differences for the order *abcd*. If any two paths turn out to be ultimately indistinguishable, then you are not dealing with a chiral centre. In the name, the symbol (R) or (S) is put at the beginning unless it is a salt or an ester in which the asymmetric centre occurs in the acid-moiety, when it precedes the name of the anion.

Example (1)

(S)-*sec*-Butylamine

Example (2)

Isopropyl (R)-2-methoxypropionate

In Example (2) the connectivity at the asymmetric centre is as follows:

The hierarchy must be established at each level of connectivity before exploring further. Here, at the first level the order of a and d are established but b and c are equal. However, at the second order, they are resolved as O is senior to H. (Note that =O counts as O,O for this purpose.) Note also that a change in connectivity within a series of compounds may result in a change, as between R and S, even though the configuration may remain constant for the series (see e.g. section 2.2 final note).

If there is more than one chiral centre on a particular chain or ring-system, then each is cited together with its locant and separated by a comma, e.g. $(1R, 3R, 7S)$-. Such a collection comes at the start as for a single centre. If the various centres are distributed on different chains, say, or on a ring and a chain, then the designators applying to the senior component go at the start of the name and the others as early as possible on the subsidiary compound radical.

Example (3)

One possible isomer might be (R)-3-[(R)-2-Bromo-1-methylethoxy]butyric acid.

Here the first (R) refers to the chiral centre on the butyric acid chain; the second to that on the ethoxy group.

2.2 GEOMETRICAL ISOMERISM

As before, the senior group or atom has to be established, but here each end of the double bond is considered separately. If it is placed in the plane of the paper and the senior groups at both ends project upwards or both are down below the plane, the compound is (Z); if they lie on opposite sides of the plane, however, it is (E).

Example (1)

At the right-hand end, O is senior to C. At the left-hand end, both are carbon atoms and each has the connectivity [C,H,H], however, at the next level they are [Br,H,H] and [H,H,H], respectively. The Br 'wins' and the senior groups lie on opposite sides of the plane, so it is an (E) isomer.

As with R and S, the (Z) and (E) come at the beginning of the name. If both Z/E and R/S designators are used in the same name, the Z or E comes first, separately enclosed inside parentheses and followed by a hyphen.

Example (2)

(Z)-(R)-1-Methyl-3-phenylprop-2-enyl acetate

Thus, the full name should be determined and the stereo-descriptors added afterwards in their appropriate places.

Example (3)

(S)-*sec*-Butyl $(1R,3R,4S)$-3-[(E)-3,4-dibromobut-2-enyl]-4-[(Z)-(R)-5,6-dihydroxy-2,5-dimethylhex-2-enyl]cyclopentanecarboxylate

Example (3) illustrates the following points:

(i) **For an ester** (see e.g. section 25.8), the name is split into two parts: radical and anion, separated by a space.

(ii) Stereo-symbols are enclosed inside parentheses. When referring to centres on attached groups, they precede the names of those groups.

(iii) Locants are supplied when there is more than one stereo-centre on the unit to be named, whether it is the main structure or an attached radical, but not when there is only one stereo-centre, e.g. the 5-position of the hexenyl chain.

(iv) The three chiral centres on the ring are at positions 1,3 and 4. For 1, (a) and (d) are easily decided but the other two neighbours (2 and 5) are each $-CH_2-$. Further exploration is needed as follows (the direction must be borne in mind throughout each exploration):

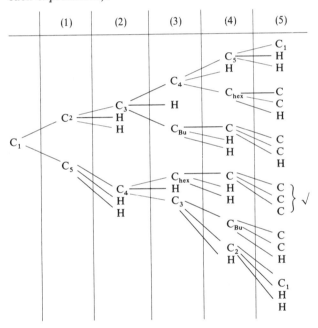

Thus five levels have to be explored in order to decide that the arm C_1-C_5 is (b) and C_1-C_2 is (c) at C_1. The C....H is conveniently shown, so no rotation of the printed structure is needed and the $a \rightarrow b \rightarrow c$ progress is clearly clockwise; hence (1R). Similar procedures at C_3 and C_4 give the full name shown alongside the structure.

(v) Ring-numbering: Position 1 is determined by the site of the carboxylate group. Direction around the ring is not resolved by the pattern of substitution (3,4—both ways) but by alphabetical preference of 'dib' to 'dih' for position 3.

(vi) Order of citation of prefixes: dib precedes dih.

Note that, in deciding the chirality-symbol at C(1), it is irrelevant that one step further would have reached the superior connectivity of $-Br$, since such jumping ahead of the stage reached is not allowed. However, for the chirality at C(3), it is the $-Br$ which is ultimately decisive in comparing the C_3-C_{Bu} and C_3-C_2 paths, viz. CHH/CHH;

CCH/CCH; BrCC/OOO and CHH—a double bond counting as two single bonds in the outward-bound sense. (To explore beyond a double bond, count two outward-bound bonds in the usual way and, for the third, return along the double bond and halve the atomic number at its nearer end).

Note. In the case of the amino acids of sections 22.1 and 22.2.5, the stereo-configuration at the so-called α-position is indicated by the symbols D and L. They do not equate consistently with R and S but are designated as follows:

3 Silicon chains

3.1 SILICON-ONLY CHAINS†

3.1.1 Straight chains

SiH_4 is called silane. Chains of composition H_3Si—$[SiH_2]_n$—SiH_3 are called disilane, trisilane, tetrasilane, etc., according to the number of silicon atoms in the chain.

3.1.2 Branched chains

The name is based on the longest unbranched chain. Side-chains are expressed as prefixes [silyl: H_3Si–; disilanyl: H_3Si–$Si(H_2)$–, etc.].

Numbering

The main chain is numbered sequentially, starting at that terminal silicon atom which assigns the lower locant-set, at the first point of difference, to the attached groups.

Example

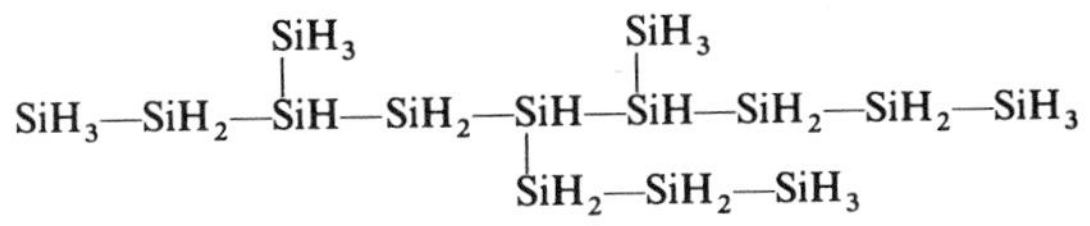

$$SiH_3-SiH_2-\overset{\overset{\displaystyle SiH_3}{|}}{SiH}-SiH_2-\overset{\overset{\displaystyle }{}}{SiH}-\overset{\overset{\displaystyle SiH_3}{|}}{SiH}-SiH_2-SiH_2-SiH_3$$
$$\overset{\displaystyle |}{SiH_2}-SiH_2-SiH_3$$

3,6-Disilyl-5-trisilanylnonasilane

[The nine-membered chain is preferred to both eight-membered chains. For the numbering 3,5,6- is preferred to 4,5,7-.]

---

† Hydrogen atoms are assumed to be present wherever necessary to satisfy the covalency of four for silicon. The hydrogen atoms are not considered as chain-members.

3.2 SILICON/OXYGEN CHAINS OF ALTERNATING ATOMS

3.2.1 Straight chains

Chains of general formula H_3Si—$[O$—$SiH_2]_n$—$OSiH_3$ are named according to the number of silicon atoms present, e.g. disiloxane ($n=0$), trisiloxane ($n=1$), tetrasiloxane ($n=2$), etc ...

Numbering

The chain is numbered sequentially, starting at that terminal silicon atom which assigns the lower locant-set, at the first point of difference, to the attached groups.

3.2.2 Branched chains

The main unbranched chain is chosen for its length and then named as in section 3.2.1; the di, tri, etc ..., prefixes signify the number of silicon atoms. For the side-chains these prefixes have the same significance. Each, as appropriate, is followed by 'siloxanyl', then 'oxy', e.g.

$$H_3Si—O—Si(H_2)—O— \qquad \text{disiloxanyloxy, etc.,}$$

As an exception, the group: H_3Si—O— is named siloxy.

Example

5,5-Disiloxypentasiloxane

Substituent groups replacing hydrogen atoms are named according to the procedures described, as appropriate, in Appendix C. Their names, in alphabetical order, precede 'siloxy' or, if attached to longer radicals or to the main chain, 'di', 'tri', etc. ... 'silanyl', or 'di', 'tri', 'tetra', 'penta', etc. ... 'siloxane', respectively. Each such group is preceded by its appropriate locant, the atoms of each main chain (both Si and O) being sequentially numbered starting from that terminal silicon atom which gives the lower locant-set for attached groups at the first point of difference.

Example

1,3,3,9-Tetrabromo-5-(bromomethoxysiloxy)-5-siloxypentasiloxane

3.3   SILICON/SULPHUR CHAINS OF ALTERNATING ATOMS

Names are constructed as for section 3.2 except that 'thia' replaces 'oxa' so that the 'a' of 'sila' is not elided. In radicals 'silylthio' replaces 'siloxy',

Example

$$H_3Si-S-SiH_2-S-SiH-S-SiH_3$$
$$|$$
$$S-SiH_3$$

3-(Silylthio)tetrasilathiane

3.4   SILICON/CARBON CHAINS

Examples (3)(7) & (8) of section 4 show situations in which isolated -Si- groups are named "silyl"; (2)(4) & (5) exemplify the use of "sila".

Where Si is the senior heteroatom from Table 3 present in the chain, the name is based on "silane" for a single Si-atom, "disilane" for two together, "trisilane" for three together, and so on. The C-groups are named according to Appendix C-1, q.v. Numbering follows the same considerations as those which apply to section 5.

Example (1) $CH_3-CH_2-CH_2-SiH_2-CH_2-CH_2-CH_3$

Dipropylsilane

Example (2)

$$\underset{2}{ClCH_2}-CHCl-\underset{}{SiH_2}-\underset{1}{SiHCl}-CH_2CH_2Cl$$

1-Chloro-1-(2-chloroethyl)-2-(1,2-dichloroethyl)disilane
(Numbering: 1,1,2 preferred to 1,2,2. Citation order: chloro, chloroe, d.)

Example (3)

$$H_3Si-O-SiH-SiH_2-SiH_2-Si(Bu)_2-SiH_3$$
$$|$$
$$OEt$$

4,4-Dibutyl-1-ethoxy-1-siloxypentasilane
(Numbering: 1,1,4,4 preferred to 2,2,5,5; citation order: b,e,s.)

4 Heterochains

The route to Box 4 implies an absence of functional groups. As Table 1 shows, many such groups can occupy chain-positions. If none of these can be identified, the answer to the Box (iv) question will be 'NO'. However, if the number of non-terminal heteroatoms of any kind in a chain is four or more, naming in terms of such functional groups may present problems. In such cases section 4 is appropriate but it should be treated as a last resort, most cases can be dealt with by recognition and use of the appropriate functional group (FG).

In particular, ester groupings should be split formally into their radical and anion moieties, the four-heteroatom rule-of-thumb being then applied separately to each (in the absence of a more senior group elsewhere on the structure) before resorting to section 4.

Example (1)

$$
\begin{array}{c}
O \\
\parallel \\
CH_3-O-P-O-CH_2-O-CH_3 \\
\mid \\
OH
\end{array}
$$

Methoxymethyl methyl hydrogen phosphate

The main chain here has seven atoms (the $=O$ and the $-OH$ are attached groups and do not count as members of the main chain), four of which are heteroatoms. However, splitting the ester group gives:

$$
CH_3-, \qquad
\begin{array}{c}
O \\
\parallel \\
-O-P-O-, \\
\mid \\
OH
\end{array}
\qquad -CH_2-O-CH_3
$$

Neither methyl nor methoxymethyl cause any difficulty in naming and there is no need to invoke the four-heteroatom criterion.

Example (2)

$$CH_3-S-NH-O-P-O-SiH_2-O-CH_3$$
$$|$$
$$OCH_3$$

5-Methoxy-2,4,6-trioxa-8-thia-7-aza-5-phospha-3-silanonane

This is a nine-membered chain having seven heteroatoms.

Although certain functional structures from Table 1 can be discerned, their use in the name would be problematical here. For example, the phosphorous acid residue cannot be used here because the atoms which flank it are not carbon. Thus it is not a phosphite ester and cannot be named as one. The section 4 procedure will instead yield a name based on 'nonane'.

The name is based on the senior chain. In a branched chain compound, this is the one which has carbon at both ends and which contains in it the maximum number of heteroatoms. If any choice remains, lowest locants are assigned to any double bonds taken as a set and thereafter the longest chain is chosen.

Numbering of the chain proceeds from that end which assigns to the heteroatoms the lower locant-set at the first point of difference. If a choice remains, number from that end which gives lower locants to heteroatoms cited earlier in Table 3 (p.270).

The name is formed by citing the hydrocarbon name from section 5 for the senior chain as if all its heteroatoms were carbons instead. This is preceded by the '-a' terms from Table 3, cited in the order of that Table; each preceded by its appropriate numeral locant. Any side-chains present are written before these '-a' terms and are expressed as radical prefixes (see Appendix C); they take their place in alphabetical order with any other substituents.

Example (3)

$$Br$$
$$|$$
$$CH_3-SiH_2-CH-CH_2-As-CH_2-CH_2-GeH_2-CH_2-SiH_3$$
$$|$$
$$CH_3-BH-CH_2-SnH_2-CH-CH_2-SiH_2-CH_3$$

8-Bromo-6-(methylsilyl)-5-(methylsilylmethyl)-12-silyl-8-arsa-11-germa-4-stanna-2-boradodecane

In this example the chain with the most heteroatoms is also the longest (twelve members), the silicon atom at potential position 13 is not counted because it would then occupy a terminal position whilst not being carbon. Numbering the heteroatoms gives either 2,5,9,11 or 2,4,8,11, and so the latter is preferred. Any double bonds are signified in the name by the endings -ene, -diene, etc., immediately preceded by locants for their positions in the chain after it has been numbered as described. If the same heteroatom occurs more than once in the senior chain, it is collected in the name by di, tri, etc.

Example (4)

$$CH_3—CH=CH—SnH_2—CH_2—Se—CH_2—SiH_2—CH_2CH_2CH_3$$

6-Selena-4-sila-8-stannaundec-9-ene

Note. (i) Se is numbered 6 from both ends so, since Si is above Sn in Table 3, it gets the lower number.

(ii) '-9-ene' means that this double bond lies between positions 9 and 10, and

(iii) the 'a' of undeca is elided before 'ene', even though a locant-numeral intervenes, but it would not be elided before a consonant as in diene, triene, etc.

Example (5)

$$CH_3—CH_2—CH=N—CH_2—SiH_2—CH_2—Te—CH_2—SiH_2—CH=CH—CH_3$$

6-Tellura-10-aza-4,8-disilatrideca-2,10-diene

(6-tellura is preferred to 8-tellura)

Note. The presence of the nitrogen atom in the chain raises the question of a functional group; however, see the statement at the start of this section.

Example (6)

$$CH_3—O—O—CH_2—BH—CH=CH_2$$

Methyl vinylborylmethyl peroxide   [section 21]

Example (7)

$$CH_3CH_2—NH—NH—NH—CH_2—SiH_2—CH_2—O—CH_2CH_3$$

1-{[(Ethoxymethyl)silyl]methyl}-3-ethyltriazane   [section 22]

Example (8)

$$SiH_3—CH_2CH_2CH_2—O—CH_2—NH—\overset{\displaystyle O}{\underset{\displaystyle O}{\overset{\uparrow}{\underset{\downarrow}{S}}}}—CH_2CH_3$$

N-[(3-Silylpropoxy)methyl]ethanesulphonamide   [section 22]

[S→O and S=O are alternative representations]

Heteroatoms in side-chains

If, discounting terminal positions, there are two or more such heteroatoms in the side-chains, the replacement prefixes oxa, aza, etc. are used in them in the same way as for main chains. If they have only one heteroatom, however, non-replacement names are used (see the substituents at positions 5 and 6 of example (3) of this section).

5 Hydrocarbon chains

5.1 UNBRANCHED CHAINS

5.1.1 Saturated *(single bonds only)*

The first four members of this series are called methane, ethane, propane and butane, respectively. After them, the name is made up of a Greek numerical stem, e.g. pent(5), hex(6), hept(7), oct(8), non(9), dec(10), etc., and then the ending: 'ane'.

5.1.2 Unsaturated

Double bonds are indicated by changing the ending to 'ene' for one, 'adiene' for two, 'atriene' for three, and so on, 'ene', 'diene', etc. being preceded by appropriate positional locants (which are dropped for 2C- and 3C-chains). Only the lower of each pair is cited; "hex-2-ene" signifies a 6C-chain with a double bond connecting C_2 and C_3.

Numbering (double bonds only)

This proceeds from the end giving lower locants for the double bonds, taken as a set, at the first point of difference.

Example (1)

$$CH_3—CH{=}CH—CH_2CH_2CH_2—CH{=}CH—CH_2CH_3$$

Deca-2,7-diene   [not -3,8-].

Chains containing triple bonds are named by changing the ending of the corresponding saturated hydrocarbon name from 'ane' to 'yne', 'adiyne', atriyne', etc., according to the number of triple bonds.

Numbering (*triple bonds only*)

This follows the procedure for double bonds.

If a chain contains both double and triple bonds, the name is formed by adding to the Greek numerical stem corresponding to the total number of carbon atoms the appropriate ending for the double bond or bonds, and then that for the triple bond or bonds—each preceded by appropriate locants. The final 'e' of 'ene' is elided before the 'y' of 'yne' (see Examples (2)–(4)).

Numbering (*double and triple bonds*)

The end which gives lower numbers to the double and triple bonds taken together as a set is numbered 1. If this is the same from either end, the end which gives lower numbers to the double bond or bonds is chosen.

Example (2)

$$CH_3—CH=CH—CH≡CH$$

Pent-3-en-1-yne (not -2- ... -4-)

Example (3)

$$HC≡C—CH_2—CH=CH_2$$

Pent-1-en-4-yne

(-1- and -4- from either end; 1 is preferred for the double bond)

Example (4)

$$CH_3—CH=CH—C≡C—CH_2—CH=CH_2$$

Octa-1,6-dien-4-yne

[1,4,6 is preferred to 2,4,7]

5.2 BRANCHED CHAINS

First, the senior chain must be identified according to section B (yellow pages) criteria, omitting (1) and (5) here.

Example (1)

$$CH_2—CH=CH—C≡CH$$
$$|$$
$$CH_2=CH—CH_2CH=CH—C—CH_2—CH_2CH=CH_2$$
$$|$$
$$CH_2—(CH_2)_7CH_3$$

The four arms branching from the quaternary carbon atom contain two, two, one and no multiple bonds, respectively. The senior chain, then, is made up of the two branches with two multiple bonds each—giving a senior chain of four multiple bonds. The name is based on the 'C-count' comprising these two arms and the quaternary carbon-atom, viz. undecatrienyne, and the other arms are expressed as radical-prefixes [see Appendix C-1] on that chain.

Numbering follows the procedure of section 5.1.2, so the name is:

6-but-3-enyl-6-nonylundeca-3,7,10-trien-1-yne

[Note elision of 'e' before 'y'].

If the population of multiple bonds is the same in more than one chain, see yellow pages, section B, criteria (8), (9) and (11–16) are applied in that order until a decision is reached.

Example (2)

$$CH_3CH_2CH=CH \begin{array}{l} CH=CH-CH_3 \\ \\ CH_2-C\equiv CH \end{array}$$

3-Prop-2-ynylhepta-2,4-diene

Example (3)

$$CH_2=CH-CH=CH-CH \begin{array}{l} CH=CH-CH_2CH_3 \\ \\ CH=CH-CH_3 \end{array}$$

5-Prop-1-enylnona-1,3,6-triene

Example (4)

$$CH_2=CH \begin{array}{c} CH_3 \\ | \\ CH-CH=CH-CH \end{array} \begin{array}{c} CH_3 \\ | \\ CH-CH_3 \\ \\ CH_2CH_3 \end{array}$$

6-Ethyl-3,7-dimethylocta-1,4-diene
[not 6-isopropyl-3-methylocta-1,4-diene]
(yellow pages criterion (B-13))

5.3 SUBSTITUTED DERIVATIVES

The senior unbranched chain is chosen as the basis of the name as described in section 5.2. (branched chains cannot form the basis of a name when substituted). The sub-branches are named in radical-prefix form (see Appendix C) and take their place in the alphabetical sequence of prefixes along with other substituents on the senior chain.

The chain is chosen first and numbered afterwards, except in applying criteria B-11, 12, 14 and 16.

Numbering

The senior chain is numbered by applying the criteria of section 5.1. If, after that, a choice persists, lower locants are assigned to the substituents, considered as a set, at the

first point of difference. If there is still a choice, lowest locants are given to the prefix
cited first in the name (following alphabetical order).

Example (1)

$$CH_3-CH=C-CH_2-CH=CH-CH_3$$
$$| $$
$$ClO_2$$

3-Chlorylhepta-2,5-diene

[not 5-chloryl ...]

Example (2)

$$\quad\quad\quad\quad Br\quad\quad\quad Br$$
$$\quad\quad\quad\quad |\quad\quad\quad\quad |$$
$$CH_3CH_2CH_2-C=C-C=C-CH_2CH_2CH_3$$
$$\quad\quad\quad\quad\quad\quad | \;\;|$$
$$\quad\quad\quad\quad\quad\quad IO \;\;IO_2$$

4,7-Dibromo-5-iodosyl-6-iodyldeca-4,6-diene

[the bromine atoms are '4,7' from both ends so the position of the iodosyl decides the
numbering.]

Example (3)

$$\quad\quad\quad\quad\quad\quad\quad\quad\quad\quad\quad CH_2-CH=CH-CH_2Cl$$
$$\quad\quad\quad\quad\quad\quad Br\quad\quad\quad /$$
$$\quad\quad\quad\quad\quad\quad |\quad\quad /$$
$$CH_3-CH=CH-C-CH$$
$$\quad\quad\quad\quad\quad\quad |\quad\quad\quad\quad\quad\quad\quad CH=CH-I$$
$$\quad\quad\quad\quad\quad\quad Br\quad\quad\quad\backslash\quad\quad /$$
$$\quad\quad\quad\quad\quad\quad\quad\quad\quad CH=C$$
$$\quad\quad\quad\quad\quad\quad\quad\quad\quad\quad\quad\backslash$$
$$\quad\quad\quad\quad\quad\quad\quad\quad\quad\quad\quad CH_2CHF_2$$

6,6-Dibromo-5-(4-chlorobut-2-enyl)-3-(2,2-difluoroethyl)-1-iodonona-1,3,7-triene

[order of prefix citation: b,c,d,i]

6 Monocycloalkanes

These are named by starting with 'cyclo' and adding the name of the open-chain hydrocarbon with the same number of carbon atoms as there are in the ring.

Example (1)

Cyclopropane

Example (2)

Cycloheptane

Example (3)

Cyclo-octane

If there are substituents attached, these are denoted in the name in the form of radical prefixes (See Appendix C) with locant numerals indicating their positions on the ring.

Numbering

This is sequential and chosen to give the lowest numbers to the substituent-set. When there is only one substituent the locant '1' may be omitted.

Example (4)

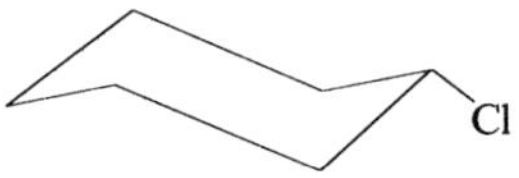

Chlorocyclohexane

Example (5)

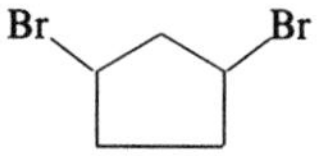

1,3-Dibromocyclopentane
[not 1,4-]

Example (6)

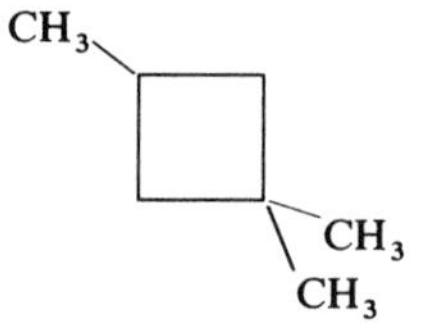

1,1,3-Trimethylcyclobutane
[not 1,3,3-]

As examples (5) and (6) show, repeated substituents are collected in the name by means of di, tri, etc.

Groups are first assigned locant-numerals as decided above and then cited as prefixes in alphabetical order before the name of the ring.

Example (7)

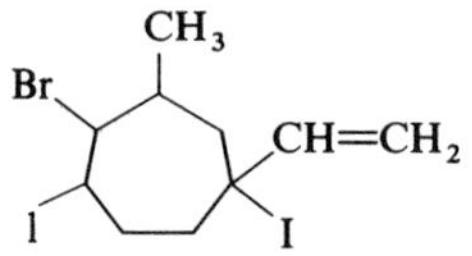

4-Bromo-1,5-di-iodo-3-methyl-1-vinylcycloheptane

Example (8)

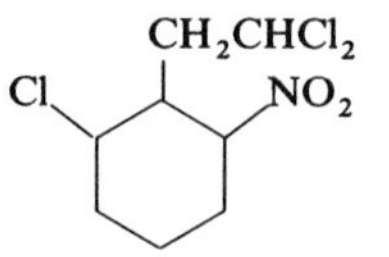

1-Chloro-2-(2,2-dichloroethyl)-3-nitrocyclohexane

Here there are two possible ring-numberings giving the substituent-set 1,2,3.

In such cases the two possible names are compared and the lower locant is preferred for the group first cited alphabetically: 1-chloro is preferred to 3-chloro.

[Note also that the compound radical inside parentheses begins with 'd'—in contrast to the di-iodo situation of example (7).]

7   Monocycloalkenes

7.1   UNSUBSTITUTED

The name for section 6 corresponding to a saturated ring of the same size is modified as follows:

— If there is one double bond, the '-ane' ending becomes '-ene'.
— For two double bonds the ending becomes '-adiene'; for three '-atriene', etc.
   If there is more than one double bond, it is assumed that the double bonds cannot occupy adjacent sides [if they do, consult CNAS].

Numbering

If all the non-cumulative double bonds that can possibly be accommodated are present (excepting aromatics of section 8, (q.v.)), or if there is only one, no locants are needed. In between these extreme conditions, locants are required and they are chosen to be as low as possible, consistent with sequential numbering around the ring.

Example (1)

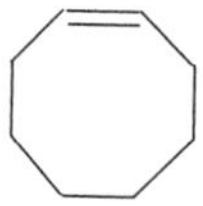

Cyclo-octene

Example (2)

Cyclonona-1,3,6-triene
[not 1,3,7 ; 1,4,6 or 1,5,7]

Example (3)

Cycloheptatriene

Example (4)

Cyclo-octatetraene

The six-membered ring with three double bonds is called 'benzene' (see section 8).

7.2 SUBSTITUTED DERIVATIVES

The attached groups are cited as prefixes, in alphabetical order when diverse, before the name of the ring (formed as above).

Numbering

This is decided by the position of the double bonds but, if that still leaves a choice (as in example (3) and (4) of section 7.1), the lowest possible locants consistent with the ring-numbering are given to the substituent-set.

Example (1)

4-Bromo-3-nitrocyclohexene
[not 5- ... -6- ...]

Example (2)

3,7,7-Trifluorocycloheptatriene
[not 4,7,7- ...]

8 Single-ring aromatics

8.1 UNSUBSTITUTED

These are named as follows:

Benzene

Toluene   o-Xylene   m-Xylene

p-Xylene   Styrene

8.2. SUBSTITUTED DERIVATIVES

Names are based on the appropriate parent compound shown in section 8.1, any additional attached groups being cited as radical prefixes [See Appendix C-0].
   This is subject to two provisos:

(i) If an alkyl chain is directly attached to any of the side-chains shown on this set of structures, the name must be based on the simpler member of the set, as appropriate.

Example (1)

$$C_6H_5CH_2O\text{—}SiH_3$$

is α-siloxytoluene rather than (siloxymethyl)benzene

Example (2)

However $$C_6H_5CH_2CH_3$$

is ethylbenzene and not α-methyltoluene.

Example (3)

Similarly

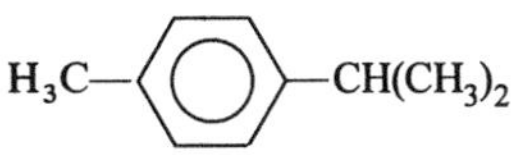

is 4-isopropyltoluene and not α,α-dimethyl-*p*-xylene.

(ii) If the ring is substituted with a group or groups already present in the trivially named parent structure shown in section 8.1, the name must be based on benzene.

Example (4)

2-methyl-*p*-xylene must instead be named 1,2,4-trimethylbenzene

Example (5)

m-vinylstyrene must instead be named 1,3-divinylbenzene.

Numbering

Locants are shown on the structures of section 8.1 but, if a choice should remain, the same numbering preferences apply as for section 6.

Example (6)

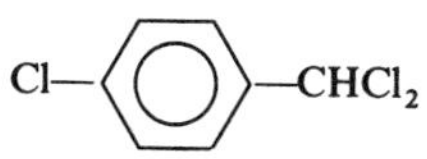

α,α,4-Trichlorotoluene

Example (7)

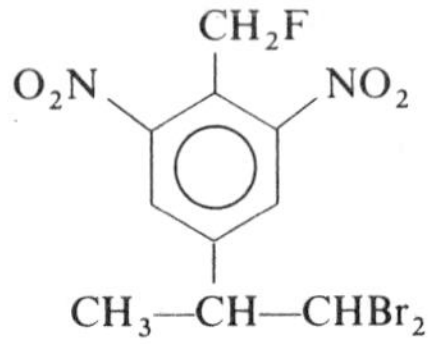

4-(2,2-Dibromo-1-methylethyl)-α-fluoro-2,6-dinitrotoluene
(not 4-[1-(dibromomethyl)ethyl]-. . . cf. B(13) Yellow pp.)

Example (8)

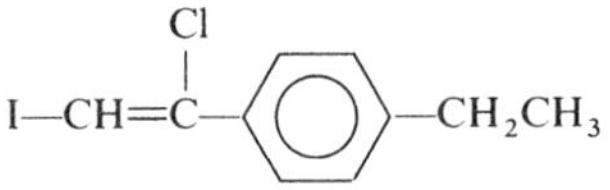

α-Chloro-4-ethyl-β-iodostyrene

9 Heteromonocycles

For a consideration of two heteromonocycles directly linked, see section 9.7.

Note that structures such as (19) and (20) of section 9.1, and (3) of section 9.3 are included here. Although chemically they are bases, they are regarded for nomenclature purposes as having no FGs in the sense of section 1.5.

BOX 9 REACHED BY ROUTE: (v)–(viii)–(ix)

(If Box 9 is reached instead by the route (viii)–(xiv)–(xx)–(xxiii)–(xxiv) see section 9.7)

9.1 TRIVIALS

If any of the following are present, their name can form the basis of the name of a substituted derivative:

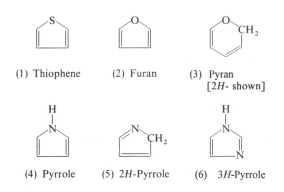

(1) Thiophene   (2) Furan   (3) Pyran
                               [2H- shown]

(4) Pyrrole   (5) 2H-Pyrrole   (6) 3H-Pyrrole

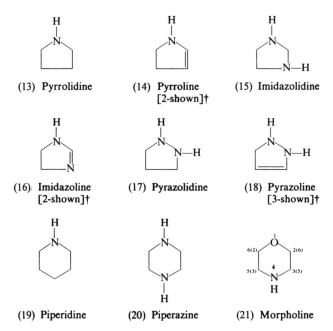

(7) Imidazole   (8) Pyrazole   (9) Pyridine

(10) Pyrazine   (11) Pyrimidine   (12) Pyridazine

In the case of hydrogenated forms, the same names for rings (1)–(12) inclusive, are retained with appropriate hydro-prefixes. However, structures (13)–(21) are named as shown.

(13) Pyrrolidine   (14) Pyrroline   (15) Imidazolidine
                     [2-shown]†

(16) Imidazoline   (17) Pyrazolidine   (18) Pyrazoline
     [2-shown]†   [3-shown]†

(19) Piperidine   (20) Piperazine   (21) Morpholine

Other single rings are named in the following sub-sections.

9.2 OTHER SATURATED† SMALL RINGS CONTAINING N

With the exception of structures (13), (15) and (17) of section 9.1 whose names are retained, if the ring is one of three, four, or five members and contains nitrogen atoms but no double bonds, its name is formed from the '-a' terms of Table 3 in the order of citation given there, with a locant for each, eliding the 'a' when a vowel follows. The name-endings are 'iridine' for a three-membered ring, 'etidine' for a four-membered ring, and 'olidine' for a five-membered ring.

† The numeral denotes the position of the double bond.

The numbering for these locants is as low as possible, consistent with sequential numbering around the ring, taking the heteroatoms as a set (a single heteroatom carries the number 1). If there is a choice, the number 1 should be given to the element occurring earliest in Table 3 and thereafter—should any choice still remain—atoms shown higher up Table 3 are preferred.

Example (1)

Aziridine
[there is no need to write 1-aziridine]

Example (2)

1,3-Diazetidine

Example (3)

1,4,2-Oxazaphospholidine
[1,2,4 is lower at the first point of difference than 1,3,4. Then O, being higher in Table 3 than P, claims position 1.]

9.3 OTHER SINGLE-RING HETEROCYCLES OF THREE TO TEN ATOMS

Provided that they are not already covered by sections 9.1 or 9.2 these are named by attaching the appropriate Table 3 terms for the heteroatoms present in the ring to one of the following endings according to (i) the ring-size, and (ii) whether it is saturated‡ or not.

When more than one kind of heteroatom is present in the ring, they are cited in the order given in Table 3 (see example (3) of section 9.2).

‡ Defined in section 1.5.

Ring-size	Saturated	Mancunide †
3	-irane	-irene
4	-etane	-ete
5	-olane	-ole
6	-inane	-ine
	[unless the Table 3 terms for O, S, Se, Te, Bi or Hg are to be cited immediately before the ending, in which case it is -ane.]	[unless the Table 3 terms for B, P, As or Sb are to be cited immediately before the ending, in which case it is -inine.]
7	-epane	-epine
8	-ocane	-ocine
9	-onane	-onine
10	-ecane	-ecine

Example (1)

N═╮
  │
  ╰═N

1,3-Diazete

Example (2)

1,2,3-Siladiborolane

Example (3)

Azepane

Example (4)

2*H*-Phosphole
[The 2*H* is necessary here to fix the double bonds.]

9.4   PARTIALLY HYDROGENATED FORMS

Apart from those forms covered in section 9.1, the unsaturated name is preceded by 'dihydro', 'tetrahydro', etc., prefixes with any locants necessary to fix the remaining unsaturation.

† Defined in section 1.5.

Example (1)

2,5,6,9-Tetrahydro-1*H*-phosphonine

The hydrogenated positions are 1,2,5,6 and 9. When there is an odd number (in this case, 5), the lowest is assigned to the indicated hydrogen, which is cited in the name immediately before that of the unsaturated ring.

When a –CH_2– group lies between two divalent ring-atoms (such as O or S), it is assumed to be hydrogenated without any recognition in the name.

9.5   RINGS OF MORE THAN TEN ATOMS

The Table 3 terms are cited as appropriate in 'Table 3-order'. Their locants are chosen by the same principles as in section 9.2, and then follows the cycloalkane or cycloalkene name (but without numbering) from sections 6 or 7 which would apply if all the heteroatoms were carbon atoms instead.

Numbering gives position 1 to the heteroatom nearest the top of Table 3 and the lowest locants thereafter are assigned to the heteroatom-set. If a choice still remains, any double bonds are then numbered as low as possible.

Example (1)

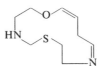

1-Oxa-6-thia-4,9-diazacyclotrideca-9,12-diene
(based on cyclotridecadiene from the rules in section 7).

However, note that the 1,4-numbering for the cycloalkene double bonds becomes -9,12- here.

Example (2)

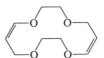

1,4,8,11-Tetraoxacyclotetradeca-5,12-diene

9.6   SUBSTITUTED DERIVATIVES

The procedures of Appendix C-0 (q.v.) are followed. If ring-numbering criteria leave any choice, lowest locants are given to the substituent-set (including any hydro-prefixes) and, if there are two or more equivalent sets, the lowest locants are given to that cited earliest in the name.

Example (1)

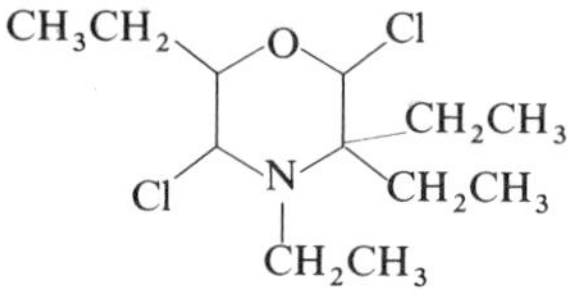

2,5-Dichloro-3,3,4,6-tetraethylmorpholine
[2,3,3,4,5,6 preferred to 2,3,4,5,5,6]

Example (2)

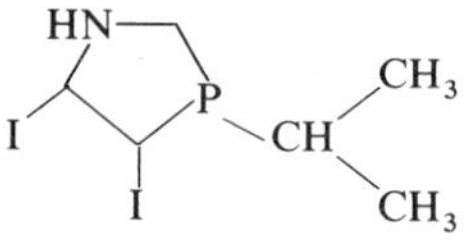

4,5-Di-iodo-3-isopropyl-1,3-azaphospholidine

Example (3)

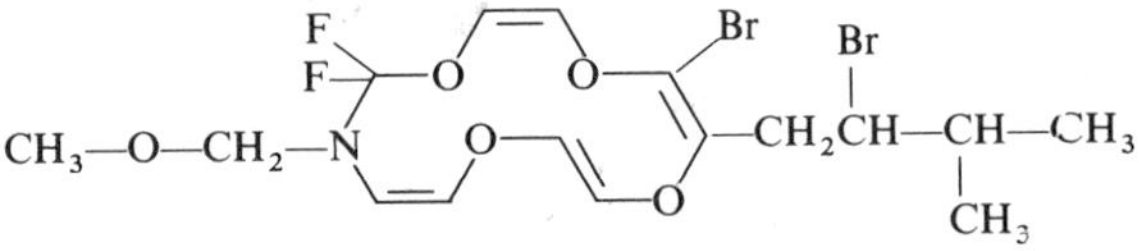

11-Bromo-10-(2-bromo-3-methylbutyl)-2,2-difluoro-3-methoxymethyl-1,6,9,12-tetraoxa-3-azatetradeca-4,7,10,13-tetraene
[1,3,6,9,12 for the heteroatoms preferred to 1,4,7,10,13 or 1,4,6,9,12]

9.7   BOX 9 REACHED BY ROUTE (viii)–(xiv)–(xx)–(xxiii)–(xxiv)

If there are two identical heterorings directly joined (disregarding substituents), the name is based on the appropriate single-ring name generated as in sections 9.1–9.5 but preceded by the locants for their positions of attachment, a hyphen, and then 'bi'. Locants for the second ring are indicated by primes.

Example (1)

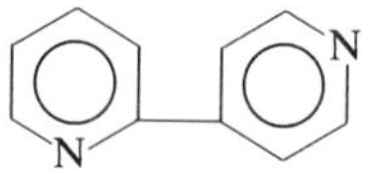

2,4'-Bipyridine
[2,4' is lower than 2',4]

Example (2)

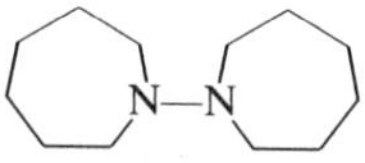

1,1'-Biazepane

If there are diverse rings, or identical rings not directly joined, the name is based on the senior ring with the other rings expressed as substituent radical prefixes (see Appendix C). The senior ring is decided by application of the criteria of section A, yellow pages.

The name of the senior ring or ring-system present is found from sections 9.1–9.5. This is cited as the name-ending; the names of any attached groups precede it in alphabetical order—with an appropriate locant (according to the numbering of the ring-system) and a hyphen immediately before each (see Appendix C as appropriate, and also Appendix D).

Example (3)

2-(2-Chlorovinyl)-5-(4,6-dimethyl-2*H*-pyran-2-yl)-2-(1,3-dioxol-4-ylmethylselenomethyl)-1,2-dihydropyrimidine
[prefix-order: c,dim,dio,h]
Pyrimidine is preferred over dioxole by criterion (A-4), yellow pages.

Example (4)

7 -(1,3,5-Triazin-2-ylmethyl)-2*H*-azepine
Azepine preferred by criterion (A-7), yellow pages.

Example (5)

3-[2-(4-Bromo-4*H*-imidazol-2-yl)ethyl]-1,4,2,3-oxathiazaphospholidine
Criterion (A-8), yellow pages, decides.

Example (6)

2-Phenyl-6-[3-(4-piperidino-1,3-dioxolan-2-yl)propyl]pyridine
(citation order : ph,pi). The criterion (A-17), yellow pages, decides seniority.

10 Monocyclic hydrocarbons with attached chain(s)

The senior chain present should be compared with the ring-system. (See the yellow pages: A, B & C.)
If there are more atoms in the senior chain than in the ring, the name is based on the chain name of section 5, with the ring expressed as a radical prefix. If not, the name is based on the ring name from section 6, 7 or 8 and the chain is expressed as a radical prefix.

If the longest chain present has the same number of atoms as the ring, the name is based on whichever has the greater number of substituents. If there is still a choice on this basis, the ring wins.

Example (1)

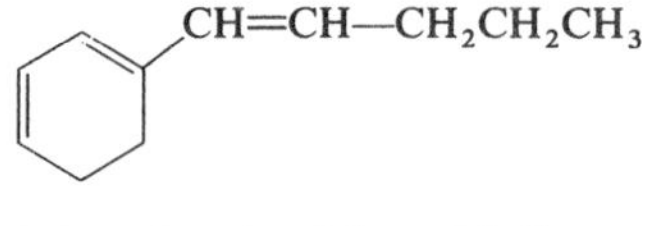

$$CH=CH-CH_2CH_2CH_3$$

1-Pent-1-enylcyclohexa-1,3-diene
[yellow pages, Criterion C-3]

Example (2)

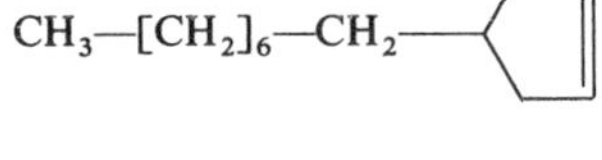

$$CH_3-[CH_2]_6-CH_2-$$

1-Cyclopent-3-enyloctane
(for numbering, see Appendix C–1.1.1 and 3.1.1)

Example (3)

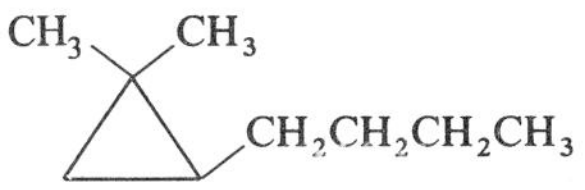

1-(2,2-Dimethylcyclopropyl)butane

senior chain: butyl (4C) vs. cyclopropyl (3C), therefore chain 'wins'.

Example (4)

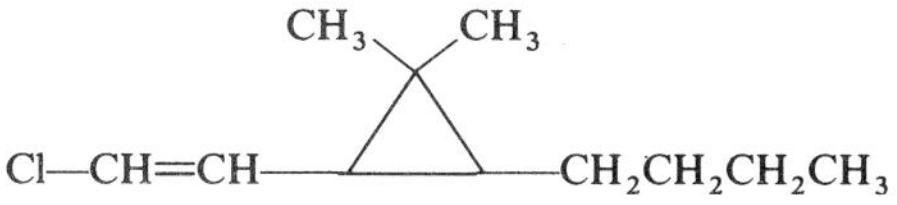

2-Butyl-3-(2-chlorovinyl)-1,1-dimethylcyclopropane

(prefix citation order: b,c,m.) Senior chain: vinyl (2C) vs. cyclopropane (3C), therefore ring 'wins' despite the presence of a butyl chain.

Example (5)

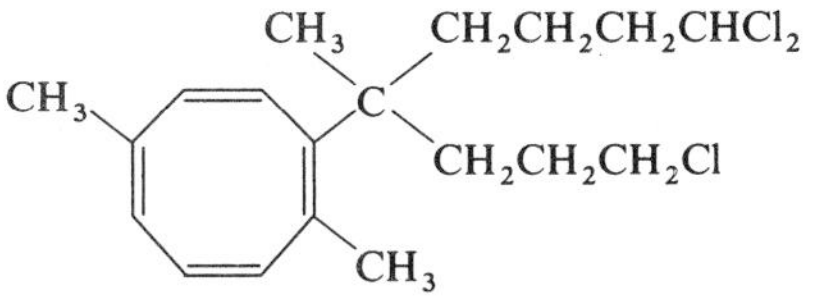

1,1,8-Trichloro-5-(2,6-dimethylcyclo-octatetraenyl)-5-methyloctane
[Senior chain and ring each have 8C; the ring has three substituents, but the chain has five substituents, therefore the chain is chosen.]

11 Bicyclic spiro-hydrocarbons

11.1 SATURATED

The name is formed by placing 'spiro' before the name of the section 5 hydrocarbon with the same number of carbon atoms. These two parts of the name are separated by a pair of square brackets containing, in ascending order, the number of carbon atoms in the two bridges, separated by a full stop.

Example (1)

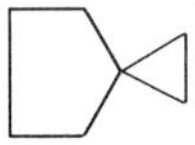

Spiro[2.4]heptane

Example (2)

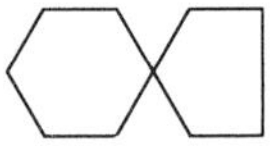

Spiro[4.5]decane

Numbering begins on the atom adjacent to the spiro-atom in the smaller ring (or, if both rings are the same size, at any of the four adjacent positions to the spiro-atom) and proceeds around the smaller ring to the spiro-atom and then on around the larger ring. The 3D representation of this situation in the Introduction [section 1.5 ('spiro-compounds')] shows that both routes around the second ring are equivalent.

11.2　UNSATURATED

Double bonds are given numbers as low as possible consistent with the constraints imposed by the rules of section 11.1.

Example

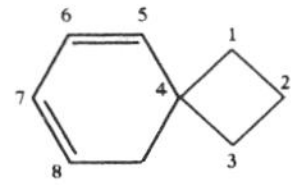

Spiro[3.5]nona-5,7-diene
(not -6,8-)

11.3　SUBSTITUTED DERIVATIVES

For the assignment of locant-numerals to substituent groups, expressed as prefixes (see Appendix C-0), the considerations in sections 11.1 and 11.2 governing numbering are followed. Where there remains a choice for numbering, lowest locants at the first point of difference are given to substituents, taken as a set, consistent with these considerations.

Example

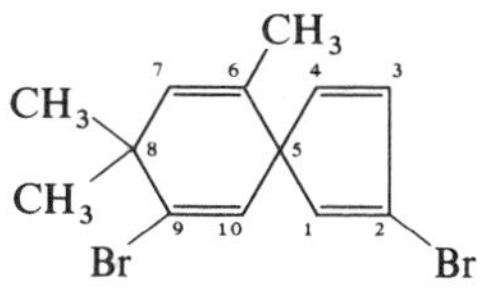

2,9-Dibromo-6,8,8-trimethylspiro[4.5]deca-1,3,6,9-tetraene

12 Two-ring aromatic† hydrocarbons

12.0 INTRODUCTION

For an aromatic structure† having two or more ring systems, none of which are in close association†, the name is based on that of the senior ring system or chain as determined by the applicable criteria of the yellow pages (q.v.) and is formed according to the appropriate section (5, 6, 7, 8, 9, 10, 12, 13 or 18.4).

12.1 FUSED RINGS

If both rings are benzene rings the name is based on naphthalene:

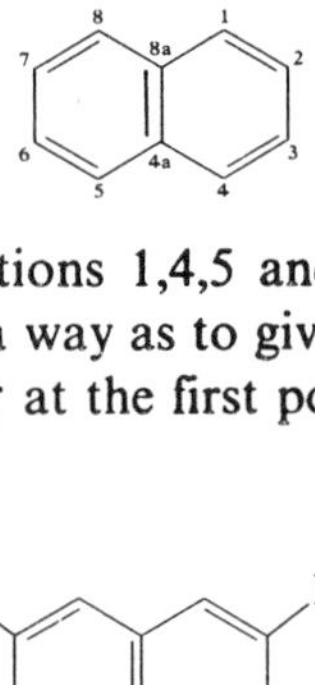

Numbering is as shown but positions 1,4,5 and 8 are equivalent and substituted derivatives are numbered in such a way as to give the substituent group names from Appendix C the lowest locant-set† at the first point of difference.

Example (1)

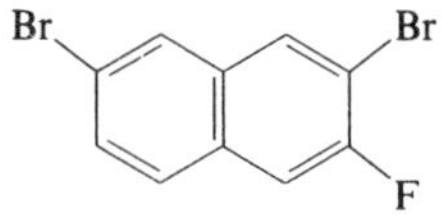

3,6-Dibromo-2-fluoronaphthalene
[not 2,7-dibromo-3-fluoronaphthalene. 2,3,6 is 'lower' than 2,3,7. The order of citation is alphabetical.]

† Defined in section 1.5.

If one ring has five members, the name is based either on

Indene   Indan
[1*H* implicit]

For each other indene isomer the saturated site is denoted in the name by an indicated hydrogen symbol (*H*) preceded by the appropriate locant, e.g. 3a*H*-indene.

If indan is further hydrogenated, the name is based on indene, unless there are less than two double bonds present, in which case, the compound is named as a bicyclononane (or -ene) under the rules of section 13.3.

Other cases are named by following 'benzo' with the name of the other ring having only one double bond, as given in section 7, except that, in such fusion-names, the ending '-ene' conveys the presence of the maximum number of non-cumulative† double bonds.

Example (2)

Benzocyclo-octene

Example (3)

5*H*-Benzocycloheptene

Note the numbering, which starts at a position on the benzene ring adjacent to the fusion-side. If a mancunide† structure has a $-CH_2-$ position, it is given the lowest possible number consistent with the above constraint on ring-numbering. [In example (3), 5*H* is lower than 9*H*.] If any choice still remains after these considerations are satisfied, then the lowest numbering is given to substitution-sites, taken as a set (including dihydro, etc.).

Example (4)

2-Bromomethyl-8,9,10,11-tetrahydro-5,5-dimethyl-5*H*-benzocyclononene

† What is defined in section 1.5 is 'cumulative'. For 'non-cumulative' *see under* 'Mancunide'.

12.2 BIPHENYL DERIVATIVES

Note that both rings are aromatic. If one ring is wholly or partially hydrogenated, the name is based on the other ring (benzene); go instead to section 8. If both rings are wholly or partially hydrogenated, the structure is named according to section 13.

The names end in 'biphenyl' and the substituents are expressed, in alphabetical order, as radical prefixes (see Appendix C).

Example

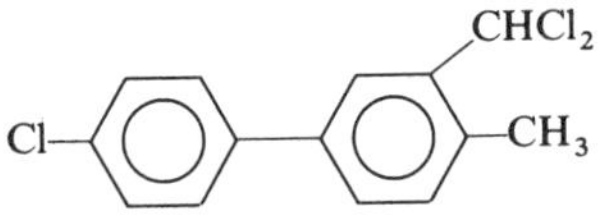

4'-Chloro-3-dichloromethyl-4-methylbiphenyl
[not α,α-dichloro-5-(4-chlorophenyl)-*o*-xylene]

12.3 ASSEMBLIES OF TWO-RING AROMATIC HYDROCARBONS

When one of the fused systems of section 12 is directly linked to the identical structure, the assembly so formed is named by prefixing the name of the fused system with 'bi', which in turn is preceded by the locants for the points of attachment on the two moieties, separated by a comma and followed by a hyphen. Locants are as low as fixed ring-numbering allows.

Example (1)

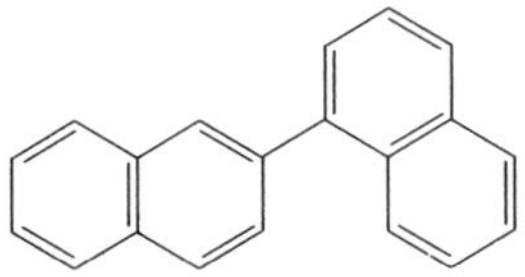

1,2'-Binaphthalene

Example (2)

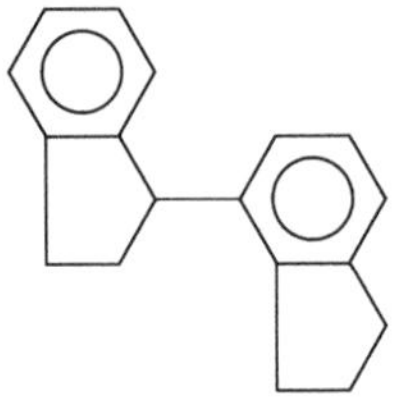

1,4'-Bi-indan
[1,4' is lower than 1',4]

13 Two-ring 'alks'

If at least one carbon-atom chain is also present, refer to section 10 before proceeding. If after that, you decide that the name is to be based on a ring or cyclic system, return here.

13.1 TWO RINGS OF EQUAL SIZE IN CLOSE ASSOCIATION

13.1.1 Rings joined by a single bond

13.1.1.1 Saturated or mancunide†

If the two rings are identically linked, '1,1'-bi' is placed before the appropriate ring-name from sections 6 or 7, enclosed in parentheses to avoid confusion with the 'bicyclo' names of section 19.1.

If the positions of attachment are different then follow the methods of section 13.1.1.2.

Example (1)

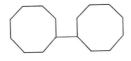

1,1'-Bi(cyclo-octane)

† Defined in section 1.5.

Example (2)

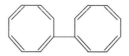

1,1'-Bi(cyclo-octatetraene)

Example (3)

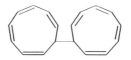

1,1'-Bi(cyclonona-2,4,6,8-tetraene)

In example (3) locants for the double bonds have been inserted in the name from section 6 since the presence of a –CH$_2$– unit in the single ring means that the ring-positions are no longer all equivalent for linking. This is the case only for odd-numbered ring-sizes.

Example (4)

1,1'-Bi(cyclopenta-2,4-diene)

Example (5)

1-(Cyclopenta-2,4-dienyl)cyclopenta-1,3-diene
Senior ring chosen by criterion A-21, yellow pages)

13.1.1.2   Partially hydrogenated forms

The name for partially hydrogenated forms is based on the least hydrogenated ring, for example cyclohexa-2,4-dienylbenzene is preferred to 5-phenylcyclohexa-1,3-diene for the following structure:

If the number of double bonds is the same on each ring and they are symmetrically linked, the name is formed according to section 13.1.1.1. If unsymmetrically linked, the one with lowest locants for the double bonds is used as a basis of the name (but, for substituted derivatives, see criterion A-21, yellow pages).

Example

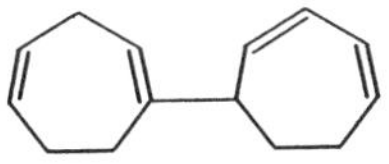

5-Cyclohepta-1,4-dienylcyclohepta-1,3-diene

13.2 TWO DISSIMILAR RINGS DIRECTLY LINKED

Follow the procedure given in section 13.4, (q.v.).

13.3 IDENTICAL OR DISSIMILAR RINGS WITH TWO† SHARED ATOMS

For the following skeletal structure:

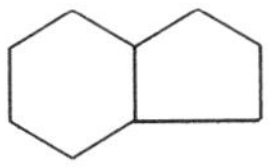

having two or more double bonds, see section 12.1, and for the following structure:

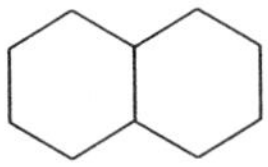

'naphthalene' is the basis of the name for all states of hydrogenation.

13.3.1 Saturated

For other rings with two shared atoms, the name of the basic structure begins 'bicyclo', followed by a pair of square brackets containing three digits separated by two full stops. These correspond to the population of the bridges linking the two shared carbon atoms (bridgeheads) cited in descending order. Finally comes the name from section 5 corresponding to the carbon-atom count for the entire ring-structure.

Example (1)

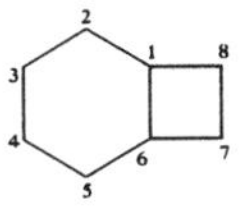

Bicyclo[4.2.0]octane

Numbering the complete structure gives 1 to the bridgehead and continues sequentially around the largest ring to the other bridgehead. It continues around the smaller ring back to position 1.

† If more than two atoms are shared, see section 19.1.

Example (2)

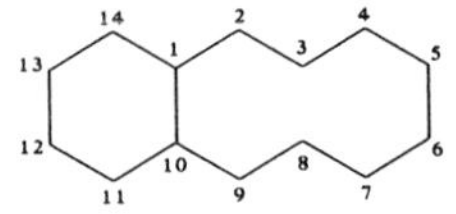

Bicyclo[8.4.0]tetradecane

13.3.2   Unsaturated

The following trivially named structures are retained as parents in naming substituted derivatives:

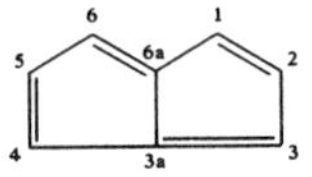

Pentalene 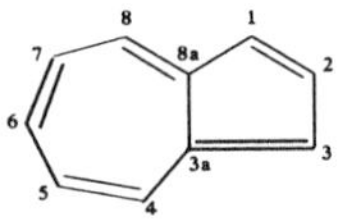   Azulene

These names are retained for hydrogenated forms by using dihydro, tetrahydro, etc. prefixes. (For fully saturated cases, see section 13.3.1.)

Numbering is as shown, any choice being resolved by assigning the lowest locant-set at the first point of difference to names for attached groups (see Appendix C) and any hydro-prefixes, all considered together.

For other unsaturated two-ring systems, the methods of section 13.3.1 are followed, locants for double bonds, taken as a set, are given the lowest numbering compatible with fixed ring-numbering constraints.

Example (1)

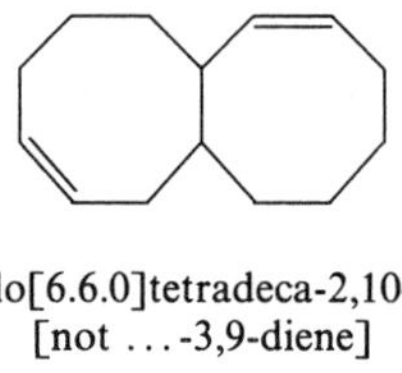

Bicyclo[6.6.0]tetradeca-2,10-diene
[not ... -3,9-diene]

Example (2)

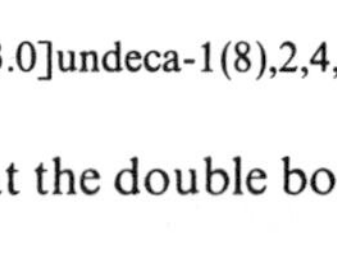

Bicyclo[6.3.0]undeca-1(8),2,4,6-tetraene

[Note: 1(*x*), where *x* ≠ 2, means that the double bond at C-1 is joined not to C-2 but to C-*x*.]

13.3.3 Substituted derivatives

For substituted derivatives the procedures of Appendix C-0 are followed. In the event of a choice arising from the ring-numbering, the lowest locants are chosen for the substituent-set. Should a choice still remain, the lowest locant is assigned to the prefix which comes earliest in the alphabet.

Example (1)

3,3-Difluoro-7-nitrosobicyclo[10.3.0]pentadec-1(12)-ene
[a of deca elided before -ene]

13.4 RINGS NOT IN CLOSE ASSOCIATION

The senior ring is chosen according to the to the criteria of section A-13, 17, 18, 20, 21, 22, yellow pages (in that order), and that ring named according to the provisions of sections 6 or 7, as appropriate. All attached groups, including other rings, are then expressed in the name in the form of radical prefixes (see Appendix C and Tables 1 and 2).

Numbering of the senior ring is determined by the unsaturation, if any. If, after that consideration, any choice for numbering remains, the set of substituents should receive the lowest possible locants at the first point of difference when the various possibilities are considered. Thereafter, the choice is determined by giving the lowest possible locant to the group cited first in the name, following alphabetical order of prefixes.

Example (1)

(Cyclopentylmethyl)cyclohexane

Without the parentheses the name would be too vague. The senior component here is decided by criterion (A-13), yellow pages.

Example (2)

1-Methyl-4-[2-(2,6,6-trimethylcyclohex-2-enyl)ethyl]cyclohexa-1,3-diene

Here the senior ring is chosen according to criterion (A-17), yellow pages. Its two substituents are at positions 1 and 4, whichever of the two sites is chosen as position 1. The issue is decided by the fact that the letter 'm' which begins 'methyl' comes before the letter 't' which begins the compound radical inside square brackets.

14 Heterospiro compounds

The name of the heterospiro compound is formed from that of the corresponding structure from section 11 (i.e., the structure with the same skeletal arrangement of carbon atoms). Heteroatoms are indicated in the name by means of the appropriate terms from Table 3 and they are cited in the order in which they appear in Table 3.

Example (1)

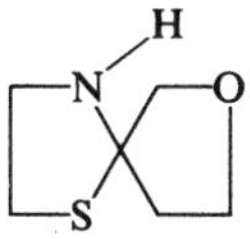

7-Oxa-1-thia-4-azaspiro[4.4]nonane

Example (2)

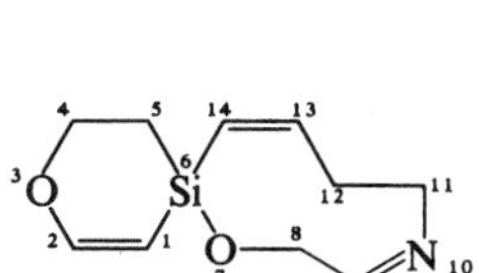

3,7-Dioxa-10-aza-6-silaspiro[5.8]tetradeca-1,9,13-triene

Any other rings present are cited in alphabetical order as radical prefixes, along with any other substituent group prefixes.

Example (3)

8-Chloromercuriomethyl-7-fluoro-8-(2-morpholinoethyl)-6,10-dioxa-2,3-diazaspiro[4.5]dec-2-ene
[Note that, in deciding the senior parent ring-system, the spiro-linked rings constitute a ring-system.]

Numbering is decided by the following criteria in the order listed:
  (i) ring-size (see section 11.1)
 (ii) lowest locants for the heteroatom set
(iii) lowest locants for unsaturation
 (iv) lowest locants for substituent prefixes (as a set)
  (v) lowest locants for prefixes cited earliest in alphabetical order.

15 Benzo/hetero two-ring systems

15.1 MANCUNIDE STRUCTURES

15.1.1 Trivially named

The following named structures are used as the basis of naming substituted derivatives using the ring-numbering shown for each:

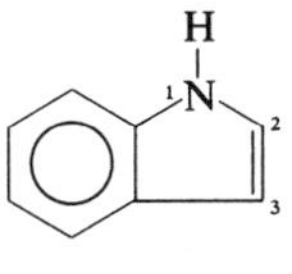

(1) Indole
[1*H*- implicit]

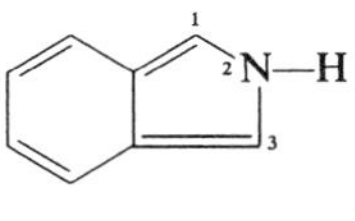

(2) Isoindole
[2*H*- implicit]

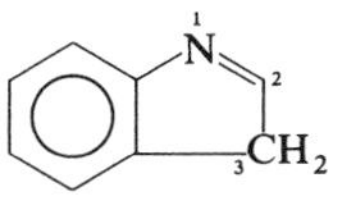

(3) 3*H*-Indole
[for example]

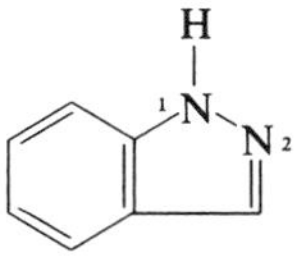

(4) 1*H*-Indazole
[for example]

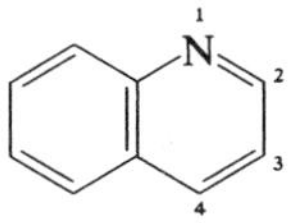

(5) Quinoline

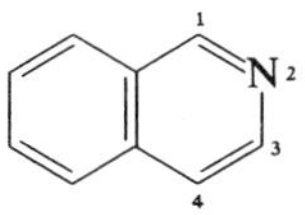

(6) Isoquinoline

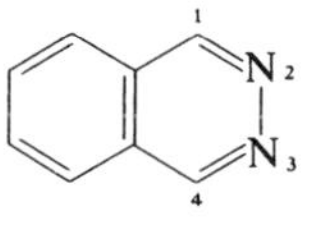

(7) Phthalazine

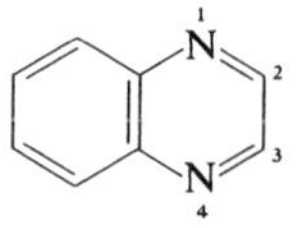

(8) Quinoxaline

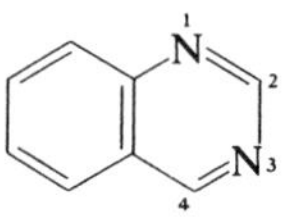

(9) Quinazoline

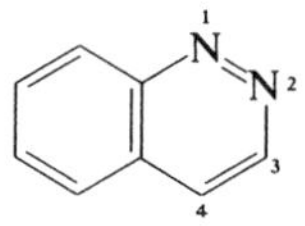

(10) Cinnoline

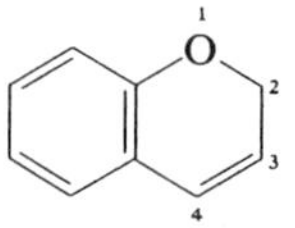

(11) 2*H*-Chromene

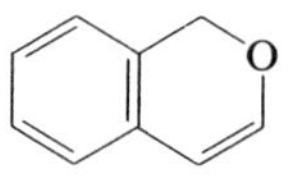

(12) 1*H*-Isochromene

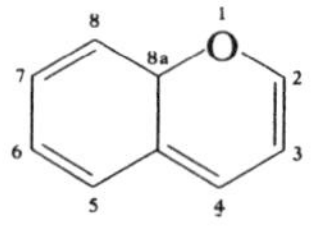

(13) 8a*H*-Chromene
[for example]

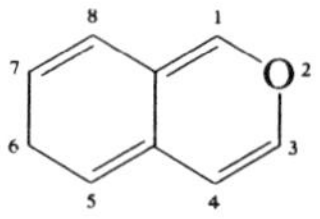

(14) 6*H*-Isochromene
[for example]

For structures (2), (7) and (8), considerations of symmetry permit an equally valid alternative numbering. In structure (2) positions 1 and 3 are equivalent; in structures (7) and (8) positions 1 and 4 are equivalent. The choice is made by the pattern of substitution.

When any of the structures (3), (4) and (8)–(12) of section 9.1 are fused to a benzene ring, the resultant two-ring system has, in each case, the trivial name corresponding to (11) and (12), and (2)–(10), respectively, of this section. The rest of those structures derived from section 9.1 are named and numbered as follows:

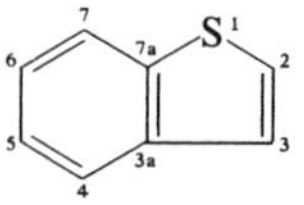

(15) Benzo[*b*]thiophene

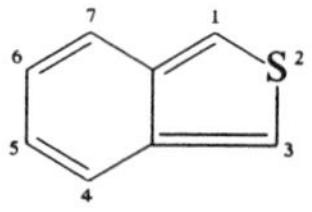

(16) Benzo[*c*]thiophene

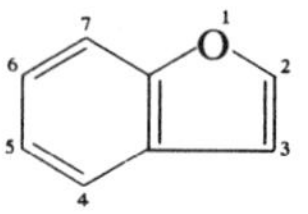

(17) Benzo[*b*]furan

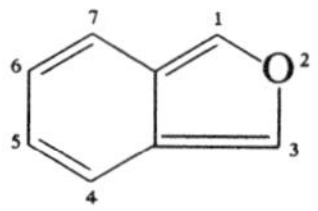

(18) Benzo[*c*]furan

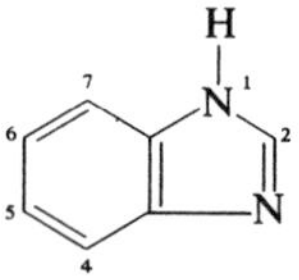

(19) Benzimidazole

Provided their hetero-ring has five or more members and less than 11, other structures are named by adding to 'benzo' the name of the appropriate single hetero-ring with the maximum number of ncdbs† taken from section 9.3.

The name is preceded by the locants for the heteroatoms in the fused system cited in the order of their occurrence in Table 3. These numbers are obtained by assigning position 1 to each of the two positions adjacent to the fusion-side in turn, numbering sequentially around the hetero-ring completely, and then comparing the two locant-sets for the heteroatoms cited in ascending order. The lower set at the first point of difference is used to precede the name of the two-ring structure. The 'o' of 'benzo' is elided before a vowel.

If the hetero-ring has only three or four members, or more than ten, then follow the procedures of section 13.3 as applied to an analogous all-carbon skeletal† structure. The positions of heteroatoms are then indicated by means of appropriate prefixes from Table 3, the locants are taken from the numbering of the carbocyclic analogue. Any choice is resolved by (i) giving the lowest locant-set to the skeletal heteroatoms, (ii) giving the lowest locants to (a) the heteroatom(s) shown earliest in Table 3, and (b) the double bond set.

Example (1)

2,1-Benzothiazole

Example (2)

1,5,3-Benzoxazaphosphepine

Example (3)

7-Oxa-8-azabicyclo[4.2.0]octa-1,3,5-triene

† Defined in section 1.5.

15.2 HYDROGENATED FORMS

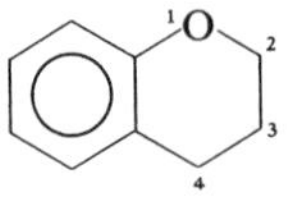

(1) Chroman

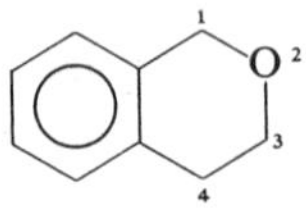

(2) Isochroman

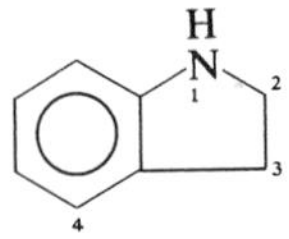

(3) Indoline

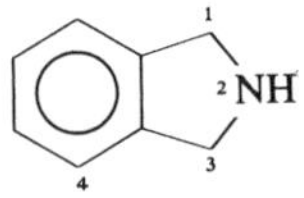

(4) Isoindoline

Structures (1)–(4) have the names shown. Their derivatives, when differently hydrogenated, are named as dihydro, tetrahydro, hexahydro, etc. derivatives of structures (11), (12), (1) and (2), respectively, of section 15.1.1.

Other hydrogenated derivatives of the structures of section 15.1.1 are named as dihydro, tetrahydro, etc. derivatives of the corresponding mancunide structures.

If there is an odd number of saturated positions, then the lowest available numbering is given to the indicated hydrogen.

Example (5)

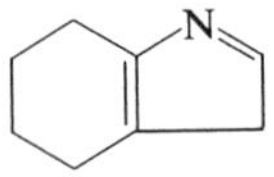

4,5,6,7-Tetrahydro-3*H*-indole

Example (6)

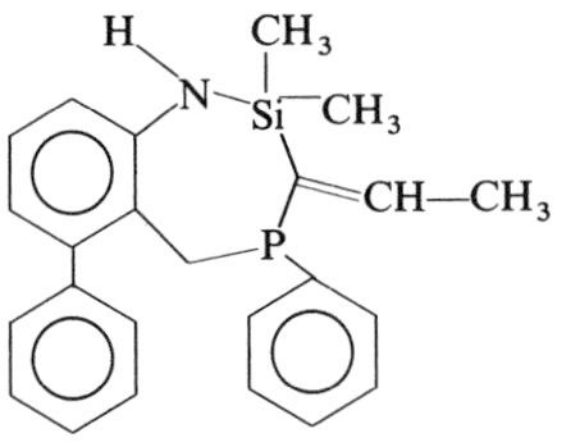

3-Ethylidene-2,3,4,5-tetrahydro-2,2-dimethyl-4,6-diphenyl-1,4,2-1*H*-benzazaphosphasilepine
[prefix citation order: e,h,m,p]
(Indicated *H* is explained in section 16.3)

Note that in example (6), 1,2,4 is lower than 2,4,5. In example (2) of section 15.1.2 both ways can give 1,3,5 but O is above N in Table 3.

16 Fused heterocycles having two rings

16.0 TRIVIALLY NAMED FORMS

The following five structures are trivially named and numbered as shown:

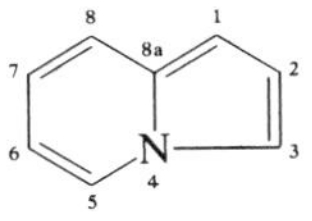

(1) Indolizine

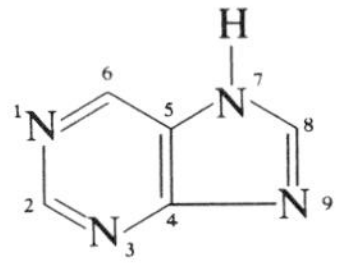

(2) Purine

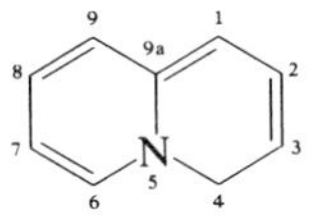

(3) Quinolizine
[4H- shown]

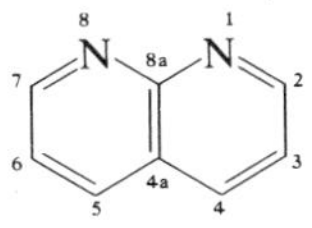

(4) Naphthyridine
[1,8- shown]

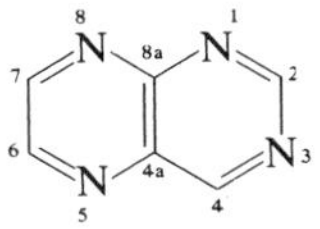

(5) Pteridine

[Name (4) may be used only when the two nitrogen atoms occur in the two separate rings as shown, and then only in non-angular positions. If either proviso is not met, the name is formed according to section 16.1.]

16.1 FUSION

This section applies to two-ring non-benzenoid heterocycles named by fusion principles. [Only two atoms are shared and no ring having fewer than five members, or more than ten, is present. If any of these conditions are not met, turn to section 19.1.]

For other cases, the fusion components (generated by splitting the structure as described below) are those having mancunide† structures, namely (9)–(12) of section 9.1 and the appropriate set from section 9.3. The hydrogenated form names cannot be used for constructing fusion names (see section 16.4).

Any hetero-ring is senior to any carbocycle. In the case of two hetero-rings, the senior ring is chosen by the criteria of section A, yellow pages, bearing in mind that shared heteroatoms are considered for this purpose as being present in each component ring, for example the following:

is regarded as being composed of and

Both rings contain nitrogen so the larger is considered senior and it is called pyrazine (see section 9.1). The smaller ring is called 1,3-oxazole by the procedures of section 9.3, and the fusion-name is formed by citing the non-senior ring-name first, but changing the final 'e' to 'o'.‡ Next, inside square brackets, come the fusion-locants (see below.) and then the name of the senior ring, thus:

[1,3]oxazolo[...]pyrazine§

Fusion locants

The sides of the senior ring are lettered *a* for 1,2; *b* for 2,3; *c* for 3,4, etc., and the fusion-side is chosen to be as early in the alphabet as the fixed ring-numbering permits. The two component rings are set side by side, with the fusion-sides lying parallel, thus:

As the arrows show, the direction of numbering of the junior ring is chosen for citation in the name so that the directions in the two rings are in the same sense along the fused side. In pyrazine each nitrogen atom could, in the first instance, be 1 and, owing to its symmetry, numbering could then go round either way. Thus, the side numbered 1,2 is chosen as *a*, 2,3 would be *b*, and so on. The numbering of 1,3-oxazole is fixed; thus the side lying next to the *a*-side of pyrazine is numbered 3,2 following the sense of the two arrows. Thus, the name of the fused system is:

[1,3]oxazolo[3.2-*a*]pyrazine
[8a*H*-isomer shown]

---

† Defined in section 1.5.
‡ As exceptions, structures (1), (2), (7), (9), and (11) of section 9.1 are called thieno, furo, imidazo, pyrido, and pyrimido, respectively.
§ For the significance of '[1,3]', see 'Indicated *H*' penultimate paragraph, section 16.3.

If one of the rings is carbocyclic, it is cited as a prefix formed from the appropriate carbocycle of section 7.1 containing *one* double bond and converting the final 'ene' to 'a'.

Then follows the italic letter for the appropriate side of the heterocycle, enclosed in square brackets, and finally the name of the heterocycle.

Example (1)

5*H*-Cyclopenta[*b*]furan
[5*H* is explained in section 16.3]

Example (2)

Cyclo-octa[*d*]pyrimidine

16.2 NUMBERING THE FUSED SYSTEM

The two rings are orientated horizontally and the number 1 is assigned to the most counter-clockwise position which is not a fusion-site (angular) on the right-hand ring.

Numbering proceeds sequentially clockwise except that angular carbon atoms do not progress in this sequence but instead add 'a' to the previous whole number.

There are four ways of doing this; the one which gives the lowest numbers to the heteroatoms, taken as a set, at the first point of difference, is chosen.

Example (1)

Structure (1) allows 1,4,5; 1,4,8; 1,5,8; and 4,5,8. Of these, the first 'wins' and it is most convenient to orientate as shown so that the clockwise sequence generates the lowest hetero-locant set.

Example (2)

Here the choice is as follows: 3,4; 4,5; 1,7; and 1,7 again.

This 1,7/1,7 choice is resolved by choosing the orientation giving the lowest locant-numerals to the atoms occurring highest in Table 3 and so the numbering and orientation shown is used.

Example (3)

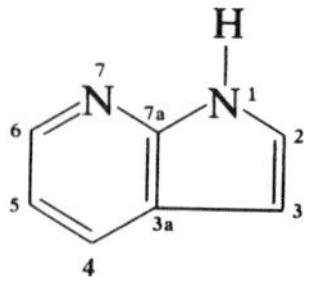

Here the choice is 3,4; 4,5; 1,7; and 1,7 again and, as Table 3 does not help, that numbering is used which combines the lowest numbering of the nitrogen atoms with lowest numbers also for the angular positions. Here, that choice lies between 3a and 4a and so the orientation and numbering shown are used.

When a heteroatom occupies an angular position it is given a full digit in the locant-sequence and not a number/letter combination as is the case when a carbon atom is so located. Thus, the first structure of section 16.1 has its oxygen atom at 1 and nitrogen atoms at positions 4 and 7; not 3a and 6.

16.3   BONDING AND INDICATED HYDROGEN

The single-ring fusion-components must be mancunide† structures (any hydrogenation is dealt with in section 16.4).

The bonding in the mancunide fused system may well be different from that in the components.

Example (1)

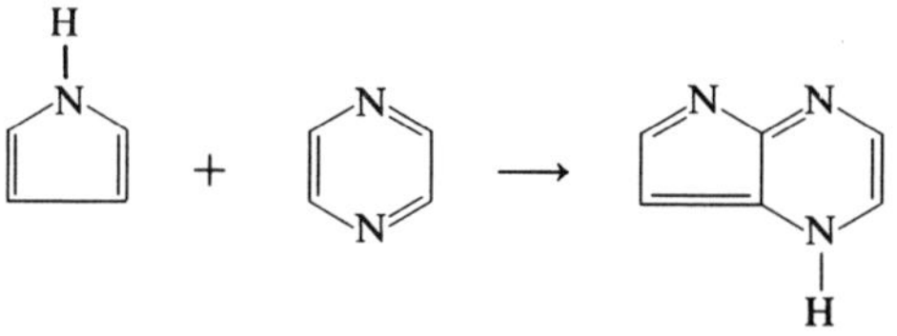

Example (2)

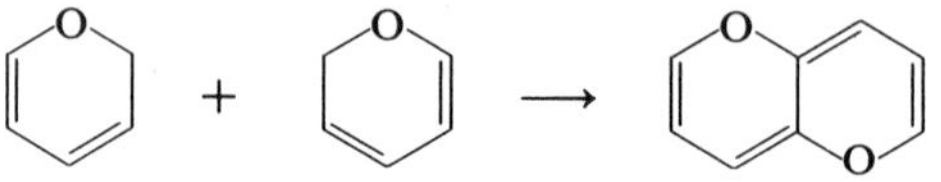

Example (3)

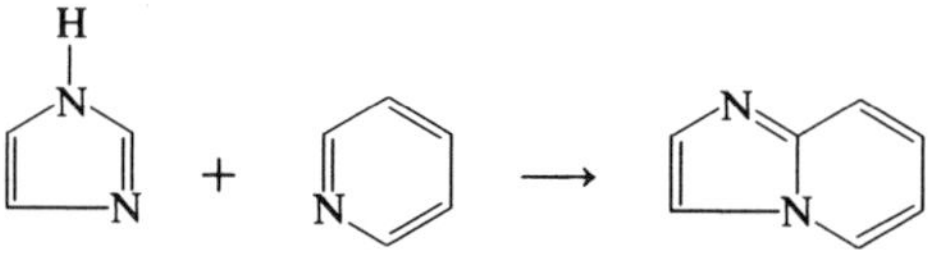

† Defined in section 1.5.

Example (4)

In example (1) components with two and three double bonds are fused to give a system with four double bonds and one saturated position. In example (2) two components with two double bonds fuse to a system with four double bonds and no saturated positions. In example (3) rings with two and three double bonds fuse to give a four double bond system with no saturated sites, and finally, in example (4) rings with two and three double bonds fuse to a system with only three double bonds; the normal valencies of 2 for O and 3 for N are maintained. However, example (4) is not a dihydro-compound. Three double bonds is the maximum that the fused system can accommodate.

Indicated hydrogen

The bonding in examples (2) and (3) is fixed but in (1) and (4) there is freedom to shift. Thus, in example (1) for instance, with the same number of double bonds the saturated position could lie anywhere.

In order that the name should apply uniquely to the isomer shown, it is necessary to fix the bonding by citing the lowest appropriate locant for the saturated position with an italic capital *H*. Thus, example (1) is named 1*H*-pyrrolo[2,3-*b*]pyrazine as the heteroatom locant-set 1,4,5 is the lowest available and position 1 is the saturation-site.

Example (2) is named pyrano[3,2-*b*]pyran and there are no saturated positions.

Example (3) is named imidazo[1,2-*a*]pyridine. Note that here the siting of the saturation on the single-ring component is not necessarily carried over into the fused system. Both nitrogen atoms of the imidazole are capable of occupying the junction position of the fused system and the fusion-locants 1,2- are preferable to the higher 3,2- for the same structure.

Example (4) is named 8*aH*-[1,3]oxazolo[3,2-*a*]pyrazine.

When one of the components of section 9.3 is involved in such fusion, its heteroatom-locants still precede the name for the component. If these numerals happen to be the same in the fused system as they were in the component, then they are merely set off by hyphens (one following if they are at the start of the name and one each side if they fall in the middle of the name). If, however, the original locants are different from those in the fused system, as in example (4), then the original locants are shown inside square brackets as shown.

Where the fixed ring-numbering constraints leave a choice for the indicated hydrogen locant, the lowest available is chosen.

Example (5)

4*H*-Cyclopenta[*c*]furan
[not 6*H*-]

16.4 HYDROGENATED FORMS

When a structure is compared with the corresponding mancunide† structure, it will be obvious whether the double bond count is less, and if so, by how much.

Where the number of non-cumulative double bonds (ncdb)† is less than the maximum possible, the name contains the prefix 'dihydro' for one double bond less, 'tetrahydro' for two less, etc., If there is an odd number of saturated (–CH$_2$–) ring positions (including e.g. –NH– but excluding such positions flanked by divalent ring atoms, e.g. O or S) the indicated hydrogen has also to be cited and it is given the lowest possible locant consistent with the ring numbering.

Example (1)

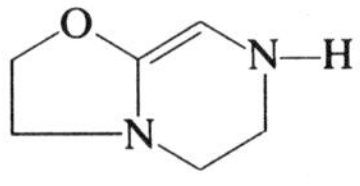

2,3,6,7-Tetrahydro-5*H*-[1,3]oxazolo[3,2-*a*]pyrazine

Note that 3,5,6,7-tetrahydro-2*H*- ... would be unrealistic here because example (4) of section 16.3 cannot be rearranged into a 2*H*- isomer.

Example (2)

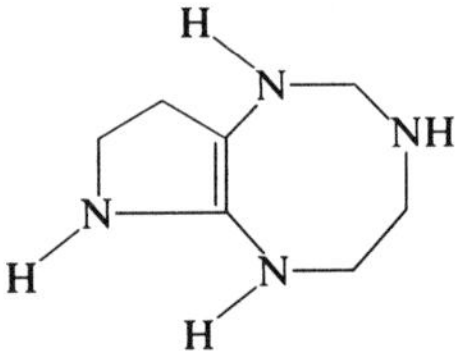

2,3,4,5,6,7,8,9-Octahydro-1*H*-pyrrolo[3,2-*d*][1,3,6]triazepine
[*d* preferred to *g*]

Compare with
Example (3)

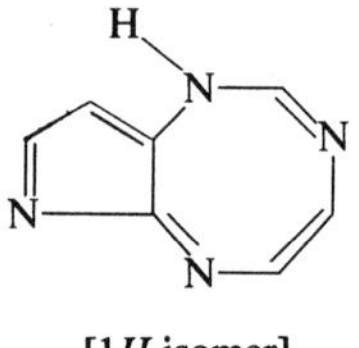

[1*H*-isomer]

Example (4)

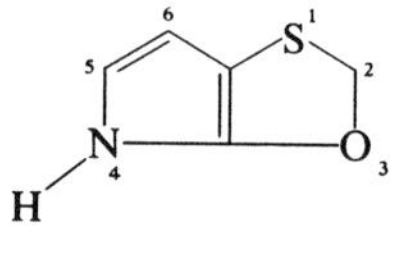

4*H*-[1,3]Oxathiolo[5,4-*b*]pyrrole
[final numbering shown].

† Defined in section 1.5.

If the structure has no double bonds the name of the corresponding mancunide†
structure is preceded by 'perhydro'. Thus, if example (1) above were to lose its only
double bond, the name would become:

Perhydro[1,3]oxazolo[3,2-*a*]pyrazine..

16.5 SUBSTITUTED DERIVATIVES

The locants for attached groups in prefix-form (see Radicals, Appendix C) are cited in
alphabetical order using the ring-numbering as decided in section 16.2. Hydro-prefixes
are included under 'h' (see also Appendix D). Then follows the name of the appropriate
parent structure derived from the instructions in sections 16.1–16.4.

Example

       (a)   (b)

To name structure (a), first remove all attached groups, then align the two rings
horizontally using only the permitted presentations (see section 17.5). Insert the
maximum number of ncdbs.†

This gives structure (b) uniquely. Reference to the section A criteria of the yellow
pages confirms that (by A-4) the smaller ring is senior. Each individual ring is to be
found in section 9; their names are pyran and pyrrole and each is shown to have a
saturated position. However, when they are fused as shown, four ncdbs can be drawn in
the positions indicated and, for a maximum, only those. The fused structure has no
tautomeric freedom and its name has no indicated hydrogen symbols, even though they
may have been necessary for the monocyclic precursors.

Numbering

With the rings arranged horizontally, either could, in principle, occupy the right-hand
position, either as shown or inverted. Thus there are four potential 1-positions and they
give the following numbers to the heteroatoms: 2,7-; 3,5-; 3,5- and 1,6-; moving
clockwise in each case. The last of these 'wins' at the first point of difference (1 is lower
than 2 or 3) and this puts the nitrogen heteroatom ring at the right with nitrogen
numbered 1.

Fusion

Pyrrole is the senior ring; pyran is to be fused to it. The name will then be of the form:
pyrano[...]pyrrole.. The sides of pyrrole, starting at 1,2=*a*, go up to *e*. Owing to
symmetry, the fused side may be *b* or *d*; *b* is the 'lower', i.e. the earlier in the alphabet.

† Defined in section 1.5.

Similarly, the pyran is numbered from the oxygen atom as 1, but one way round gives 4,5- for the fusion locant whilst the other way gives 3,4-. (This is the form in which the numerical comparison should be made, although the parallelism of sense of the matching sides (see 16.1 Fusion locants) at the fusion-site demands 5,4 rather than 4,5). The fusion locants are thus [3,4-*b*].

As structure (a) has only one double bond, it must be a hexahydro derivative of (b). Appropriate locants are 1,2,3,4,5,7. Next the names for the attached groups are added and arranged in alphabetical order:

7-Bromo-1,2,3,4,5,7-hexahydro-4,4-dimethylpyrano[3,4-*b*]pyrrole
[order of citation: b,h,m].

17 Multi-ring heterocycles

17.1 ASSEMBLIES OF THREE OR MORE IDENTICAL MONOCYCLES

The rings, bonded to each other by direct linkage, are individually named in accordance with the appropriate part of section 9. The name starts with 'ter' for three identical rings linked together, and 'quater' for four rings, but before that comes the locants for the junction-site on each ring. Those for the first ring are unprimed, those for the second primed, those for the third double-primed and so on. The junction-locants for each pair of joined rings are separated by a comma and each such pair is isolated from the next by a colon. The last locant is followed by a hyphen. Where there is any choice, the junction-locants are chosen to be as low as possible, consistent with the fixed ring-numbering.

Example (1)

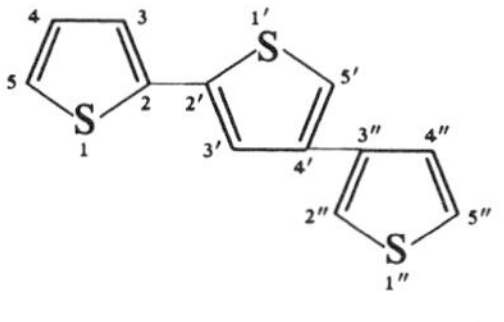

2,2′:4′,3″-Terthiophene

Example (2)

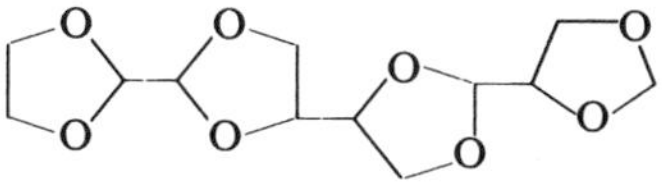

2,2′:4′,4″:2″,4‴-Quater-1,3-dioxolane

Hydrogenated forms intermediate between fully saturated and fully unsaturated are treated in a manner analogous to those of section 9.4, provided that rings grouped by ter, quater, etc. prefixes are identical. If any are differently hydrogenated, only the identical ones can be so linked in the name. In cases of diverse hydrogenation-states criterion (A-17), yellow pages, decides the senior parent.

Example (3)

2-[5-(1,4-Dihydro-2-pyridyl)-4,5-dihydro-2-pyridyl]-4-piperidinopyridine

17.2   SUBSTITUTED DERIVATIVES OF MONOCYCLE ASSEMBLIES

The procedures of Appendix C-0 are followed. Cyclic groups, even if identical with the units of the assembly, are included in this prefix-set if they are separated from the assembly by a chain, however short.

Example (1)

3'-Bromo-5"-furfuryl-2,2':4',3"-terfuran

17.3   SYSTEMS WITH THREE OR MORE RINGS FUSED TOGETHER

The same considerations apply for name construction as in section 16.1, except that trivially named two-ring and three-ring systems (such as quinoline and acridine) can also take part as fusion components, as can the two-ring benzo-heterocycles of section 15, the mancunide† carbocycles and heteromonocycles of sections 7.1, and sections 9.1 and 9.3 respectively, and the trivially named polycyclics of section 18.1.

  Trivially named systems usable as fusion components also include:

(1) Thianthrene   (2) Xanthene
                                   [9H- implicit]

† Defined in section 1.5.

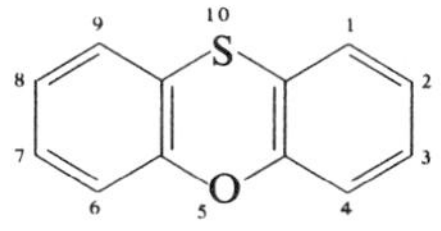

(3) Phenoxathiin

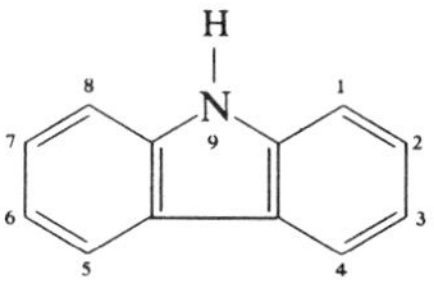

(4) 'Carbazole'
[9*H* implicit]

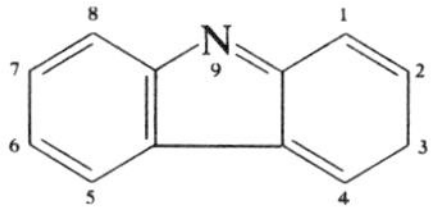

(5) 3*H*-Carbazole
[for example]

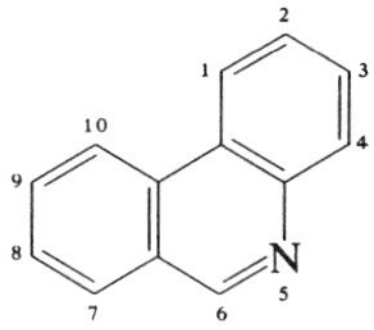

(6) Phenanthridine

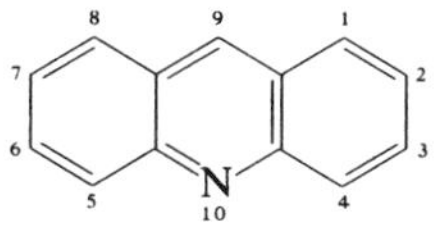

(7) Acridine

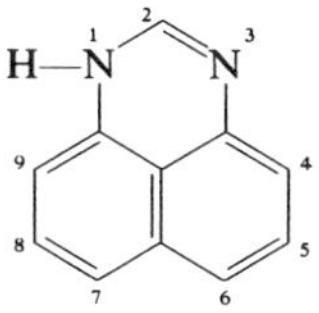

(8) Perimidine

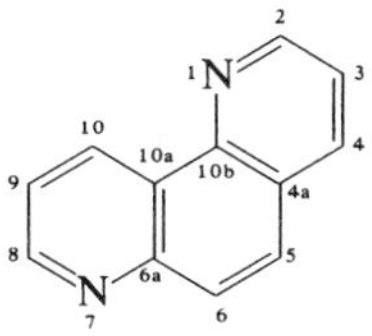

(9) Phenanthroline
[1,7- shown]

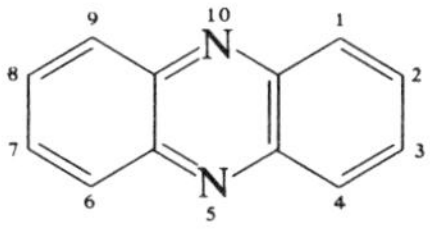

(10) Phenazine

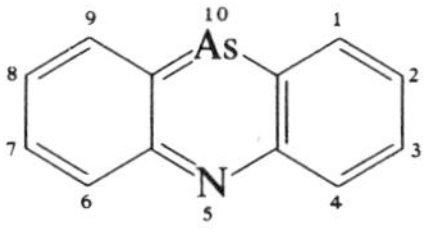

(11) Phenarsazine

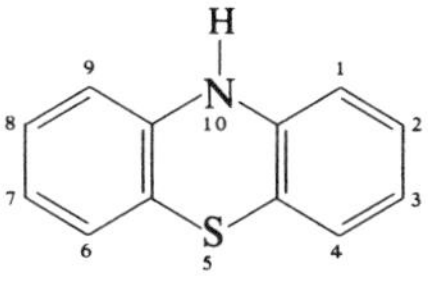

(12) Phenothiazine

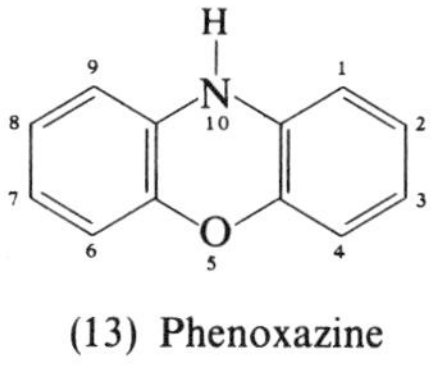

(13) Phenoxazine

As regards the citation or omission of indicated H in structures (2)(4–5)(8)(12) & (13), the remarks concerning structures (2) & (3) of section 18.1 apply (see p.80).

Name (9) may be used only when the two nitrogen atoms occur in the two separate rings as shown, and then only in non-angular positions. If these conditions are not met, the name is formed from appropriate components by fusion (see section 16.1).

Note that, with the exceptions of structures (2) and (7), these structures are numbered according to the methods of section 17.5 (q.v.). When these systems are used as fusion components, the same methods are used as in section 16, q.v. For the choice of base component, see section A, yellow pages. The fusion-name so formed applies to the structure with the maximum number of ncdbs† and it may be necessary to rearrange the bonding of the original components to achieve this.

'Normal' valency considerations are taken to apply. If fewer than this maximum number of ncdbs is present, the state of hydrogenation is conveyed by dihydro, tetrahydro, etc. prefixes cited in alphabetical order among any other substituent prefixes.

If the ring-structure of the compound can be built up by a single fusion of chosen permitted components, the name is constructed exactly as described in section 16. ['Quino' is used instead of quinolino and 'isoquino' instead of isoquinolino.]

Example (14)

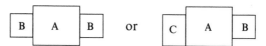

1*H*-Chromeno[2,3-*b*]quinoline

For fused polycyclic systems, the senior component structure has to be identified from the following possibilities: section 9.1, (1)–(14) and (7)–(12); mancunide† structures of section 9.3; section 15.1, (1)–(14); section 16.0, (1)–(5); and, from this section, (1)–(13), as appropriate after application of criteria A, yellow pages (q.v.).

If more than a single fusion is necessary then one of three situations has to be identified:

(i) The senior component has two other components fused to it which are not in contact with each other.

| B | A | B | or | C | A | B |

In this case, the sides of the senior component are lettered sequentially; positions 1 and 2 are linked by side *a*, and so on; and the minor components are individually numbered (using either fixed numbering or following section 16.2, as appropriate).

Example (15)

Furo[3,4-*f*]thieno[3,4-*h*]quinoline

† Defined in section 1.5.

The fusion-prefixes for the minor components are cited in alphabetical order, and the fusion-locants are denoted in the manner as described in section 16.1 (q.v.).

(ii) The senior component is fused to a minor component which is in turn fused to another minor component remote from the senior component:

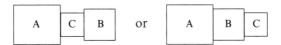

Here, the fusion between the senior and first minor component is denoted just as in section 16 but the minor component numbering is continued around its periphery and used to denote the second fusion; its sides are denoted in the fusion-locants inside another pair of square brackets, [ ], by means of the appropriate numerals. The fusion side of the third component is indicated, before a colon, by 'primed' numerals. Otherwise, everything has the same significance as in the single fusion case. The order in the name is: outermost minor component, central minor component, senior component.

Example (16)

Pyrido[1',2':1,2]imidazo[4,5-*b*]quinoxaline

If two or more adjacent edges are fused, all the fusion locants concerned are cited in the order determined by parallelity (see Section 16.1 "Fusion locants") and, in case of a choice, by the low locant-set principle.

Example (17)

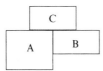

Phosphorinol[4',3',2':3,3a,4]azuleno[2,1,8a,8-*cde*]isoindole
[*cde* preferred to *ghi*; 1,2,8,8a preferred to 2,3,3a,4; 4',3',2' preferred to 4',5',6'.
Contrast order for seniority comparison (1,2,8,8a) with that for citation in the fusion name.]

(iii) The third case covers situations which cannot be treated under (i) or (ii), e.g. where three components are fused together, each in fusion contact with the other two:

In such cases, the corresponding all-carbon structure from section 18 forms the basis of the name. Locants considered as a set for the heteroatoms, named by replacement terms from Table 3, are given the lowest numbers, considered as a set, subject to the constraints of the carbocycle-numbering, which is fully retained however many heteroatoms are inserted (so long as any carbon atoms remain).

Example (18)

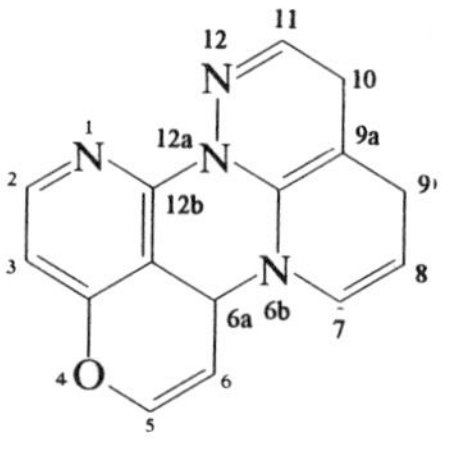

6a*H*,9*H*,10*H*-4-Oxa-1,6b,12,12a-tetra-azaperylene
(1,4,6b,12,12a preferred to 1,6a,9,12,12b)

17.4 ASSEMBLIES OF FUSED HETEROCYCLIC SYSTEMS

These assemblies consist of the heterocyclic ring-systems of section 15, 16 or 17.3 linked directly in pairs of identical components.

Names are formed by citing in order:

(a) the locants for the points of linkage, as low as fixed numbering will allow, cited in ascending order, separated by a comma and followed by a hyphen.
(b) 'bi', and
(c) the name of the fused component, as already described in sections 15–17.

Example (1)

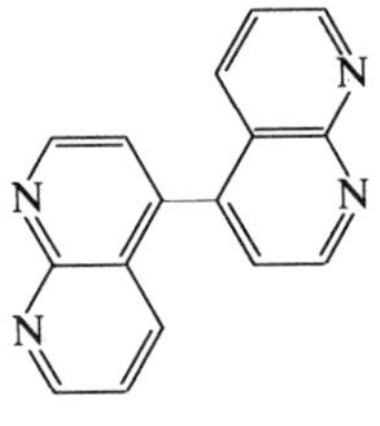

4,4'-Bi-1,8-naphthyridine
[not 4,5' nor 5,5']

Example (2)

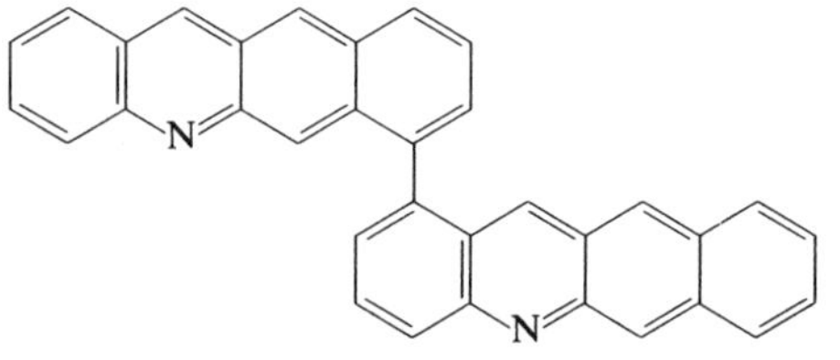

1,7'-Bibenzo[*b*]acridine

17.5 SUBSTITUTED DERIVATIVES (sections 17.1–17.4)

The procedures of Appendix C-0 (q.v.) are followed. For this purpose, the ring systems must be numbered. Before that can be done they must first be correctly orientated.

Firstly, the ring-system must be so orientated that the maximum possible number of rings lie in a horizontal line 'East–West' (the paper being considered as 'North–South' from top to bottom). For this purpose, five-membered rings may be drawn either as:

or as: but not in this form:

Next, the orientation is chosen which also gives the greatest number of rings in the 'North–East' quadrant when the horizontal ring-row is bisected by a vertical axis.

If a choice remains, the orientation which gives the lowest number of rings in the 'South–West' quadrant is preferred.

After these requirements have been satisfied, the orientation which assigns the lowest numbers to the heteroatom-set, at the first point of difference, is preferred. Any further choice is resolved by preferring the elements of Table 3 for lowest numbering in the order given there.

Next, lowest locants are assigned to any indicated hydrogen (see section 16.3). Then, lowest locants are assigned to substituents, including dihydro-, etc., prefixes, taken as a set.

Thus, in section 17.3, structures (2)–(5), (7), (11)–(13) and (14) all have two possible positions numbered 1 which would have to be decided by the pattern of substitution in their derivatives. Structures (1) and (10) have four such sites. In example (14) of that section, exchanging positions 1 and 10 would still give the locants 5 and 6 for the heteroatoms, but, as O is higher than N in Table 3, 5 is preferred to 6 for O.

When the orientation criteria are applied, in the order given, to example (15) of section 17.3, there are three ways of choosing two rings in the E–W line and each choice can then be arranged in four ways. Of these 12 possibilities six place 1½ rings in the NE-quadrant and a ½-ring in the SW quadrant.

Numbering

When the ring-system is so orientated, it is numbered sequentially clockwise starting from the most anti-clockwise atom (not involved in fusion) of the extreme right-hand uppermost ring. This procedure gives the following possible numberings for the heteroatoms in the six preferred orientations: 2,7,9; 2,5,10; 2,5,7; 2,4,9 (twice); and 1,6,9. Of these, the last is the lowest at the first point of difference and the orientation and numbering for substituent locants is:

In this case the lowest locant set was unique. In the event of a choice between two or more minimum locant-sets, priority is given to the elements occuring earliest in Table 3.

It may be that a choice remains. For example, consider section 17.3 ex.(16). It will be seen that, of the four possible orientations which give the four rings in a horizontal row, each assigns to the nitrogen atoms the locants 5,6,11,12, and in such a case, the orientation is chosen which gives the lowest numbers to the carbon atoms common to two rings. This is achieved by rotating the structure shown on p.75 about the horizontal and vertical axes in the paper-plane. This gives them the locants 4a,5a,6a,10a,11a; the alignment depicted would give 4a,5a,10a,11a,12a and the other two possibilities also give a higher locant-set.

18 Multi-ring carbocycles

18.1 TRIVIALLY NAMED RING-SYSTEMS

These are named and numbered as shown.

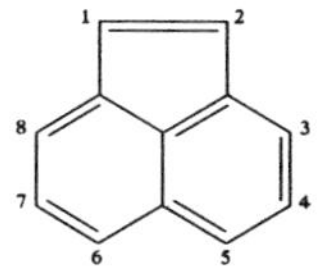

(1) Acenaphthylene

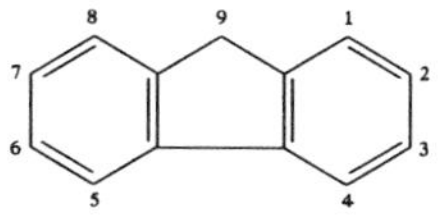

(2) Fluorene

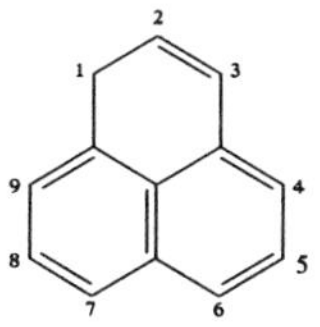

(3) Phenalene

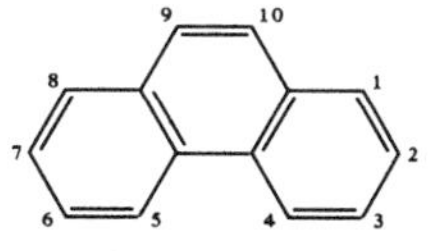

(4) Phenanthrene

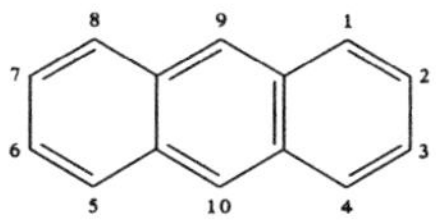

(5) Anthracene

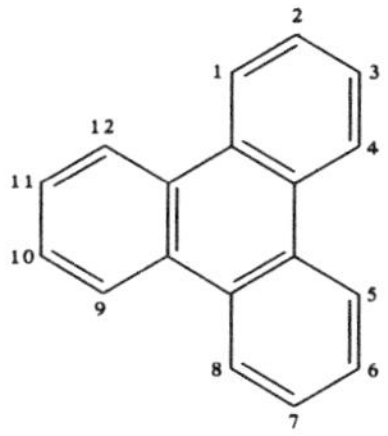

(6) Triphenylene.

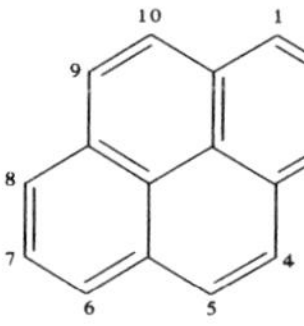

(7) Pyrene

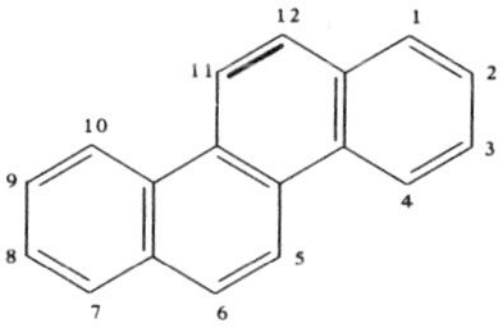

(8) Chrysene

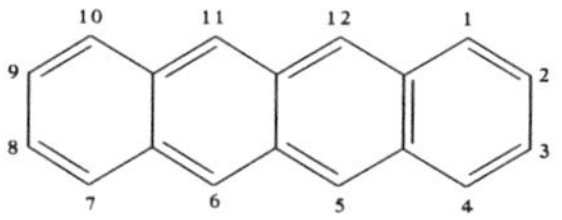

(9) Naphthacene

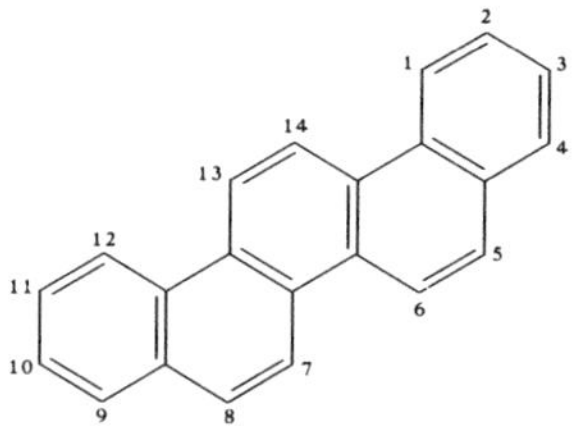

(10) Picene

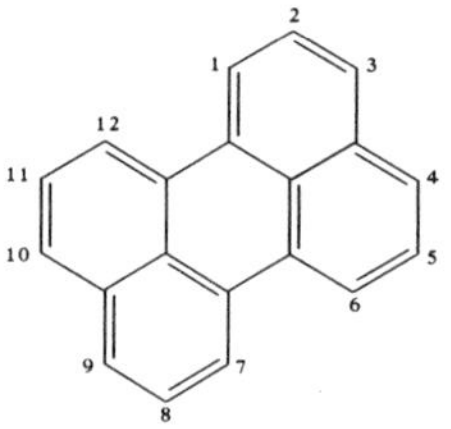

(11) Perylene

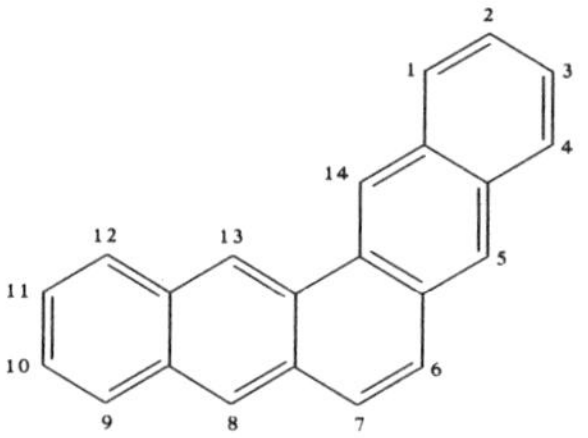

(12) Pentaphene

In the case of structures (2) and (3) the name given implies the structure shown and if the $-CH_2-$ position is shifted to another site, the same name has to be preceded by the appropriate locant and indicated hydrogen.

Example:

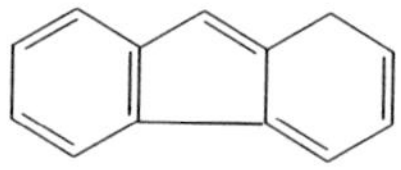

1*H*-Fluorene

Structures (4) and (5) are exceptional in not following the orientation and numbering rules given in section 17.5. Otherwise, these apply to carbocycles (except that considerations relevant to heteroatoms do not apply.)

18.2 FUSED POLYCYCLIC HYDROCARBONS

Names are based on the component occurring latest in the list of structures of section 18.1. [nb. There are a number of other trivially named structures which could have been included in the list but they are comparatively rarely encountered.]

More complicated fused structures are named by citing the prefix or prefixes for minor components before the name of the base component drawn from the list of structures in section 18.1. Such prefixes are formed by changing the 'ene' ending of the minor components to 'eno', except for the following contracted forms which are used instead: 'acenaphtho' for structure (1); 'phenanthro' for structure (4); 'anthro' for structure (5); 'perylo' for structure (11); 'benzo' for benzene, and 'naphtho' for naphthalene.

The two parts of the name are separated by a pair of square brackets which enclose the fusion-locants, small italic letters for the sides of the main component, and numbers for those of the minor components (chosen in both cases to be as low as possible whenever there is a choice). If a minor component is repeated, it is collected by means of di, tri etc., as appropriate. As stated for the examples of section 17.3 (which should be read together with this section), the electronic arrangement in components may be altered by fusion. As before, the name of the final structure implies the presence of the maximum possible number of ncdbs.† Benzene is exceptional in requiring no locants for its fusion mode as all six sides are equivalent.

Example (1)

Dibenzo[*f,pqr*]picene

It can be seen that the structure of example (1) could be built up from a variety of possible components, e.g. (3) + (6) or (4) + (7) but these are rejected because they occur earlier than (10) in the list in section 18.1. Should there be a choice even when the base-component has been decided, the minor components should be as simple as possible. Notice also that the locants [*f, pqr*] are preferred to [*j, tuv*] even though that gives the same structure; this is because *f* is 'lower' than *j*.

For naming substituted derivatives, the final fused structure has to be numbered and, as in section 17 the orientation rules of that section have first to be applied. Accordingly, when example (1) is aligned with three rings E–W (the most that can be achieved) and the maximum number of rings in the N–E-quadrant, it is numbered as shown, commencing with the most counter-clockwise atom not in a fusion-site on the most 'North-Easterly' ring, and then proceeding around the periphery in a clockwise direction, skipping the angular positions.

† Defined in section 1.5.

If any other cyclic components are attached to the structure other than by fusion, they are cited in the form of prefixes (see Appendix C) taking their place, along with any dihydro, etc., prefixes with those for any other groups in alphabetical order.

Example (2)

11,11-Dichloro-6,6a,11,11a-tetrahydro-8,9-dimethoxy-6,6-dimethyl-3-styryl-5*H*-
indeno[2,1-*a*]phenanthrene
[prefix citation order: c,h,metho, methy, s]

[Note that the fused ring-system has eight double bonds instead of the maximum possible number of ten—thus, it is a tetrahydro- compound. Of the five saturated sites the lowest locant is assigned to the indicated hydrogen. Note also that the name is not based on naphtho[*a*]fluorene because phenanthrene occurs later than fluorene in the section 18.1 list.]

To sum up, the operations, in order, are as follows:

(1) Identify the most senior principal component before fusion, consulting section A, yellow pages, if necessary,
(2) Identify the minor fusion-components and the mode of fusion; assign appropriate fusion-locants,
(3) Orientate the structure correctly,
(4) Number the structure systematically,
(5) Name the attached groups and assign to each group appropriate locants, also give appropriate locants to any hydro-prefixes,
(6) Cite their names (preceded by their locants and a hyphen) in alphabetical order before that of the fused system.

18.3 OTHER MODES OF CLOSE ASSOCIATION

If three or more rings or ring-systems are directly linked, the following cases may occur:

18.3.1   All identical structures

(Although not necessarily identically substituted).

The name is formed by starting with 'ter' for three identical structures; 'quater' for four and 'quinque' for five. Then follows the name of the repeated ring or ring-system. The mode of linkage is conveyed by citing the lowest locants consistent with any fixed numbering, an end-position unit having unprimed numbering, the next primed, the next double-primed, and so on. These locants are cited in pairs (one pair for each junction) separated from each other by a comma, and from other locant-pairs by a colon.

Example (1)

1,1':3',1"-Tercyclohexane

Example (2)

2',2",4'-Tribromo-1,1':3',1":3",1"'-quatercyclobutadiene

[Where there is a choice, lowest numbering is chosen (a) for points of attachment between rings, and (b) for substituents, taken as a set. Here, 2',2",4' is 'lower' than 2',2",4".]

Exception

In this situation, benzene rings are named 'phenyl'.

Example (3)

1,1':3',1"-Terphenyl

18.3.2   The structure containing two or more identical units directly linked

Rings or ring-systems which can be collected in the name by means of 'bi' or 'ter', 'quater' etc., as in section 18.3.1, are so named. Others are expressed as substituents on the identical assembly in the form of radical prefixes (see Appendix C).

Example (1)

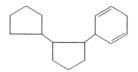

2-Cyclohexa-2,5-dienylbi(cyclopentane)

[The parentheses are used to avoid any possible confusion with a bridged-ring compound of the type covered by section 19.1.]

Exception

If the structure contains directly linked benzene rings, two together are called 'biphenyl', three are called 'terphenyl', etc.

Example (2)

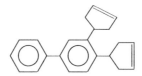

3,4-Dicyclopent-3-enylbiphenyl

If more than one kind of ring or ring-system is present, the name is based on the senior ring-system, decided by the seniority rules of section A, yellow pages, other rings being cited as appropriate radical prefixes (see Appendix C).

Lowest numbers are assigned to radical attachment positions, subject to the constraints of giving lowest locants to linkage sites of the assembly.

Example (3)

4-(1,1':3',1''-Tercyclobutan-3-yl)biphenyl

18.3.3 Assemblies of fused polycarbocycles

For linked identical systems of only two rings, see section 12.3.

These consist of the fused systems of sections 18.1 and 18.2 linked directly in pairs of identical components. Names are formed by citing in order:

(a) the locants for the points of linkage, as low as fixed ring numbering allows, cited in ascending order and separated by a comma and followed by a hyphen,
(b) 'bi', and
(c) the name of the fused component, formed as described already in section 18.1 or 18.2, as appropriate.

Any indicated hydrogen (see section 16.3) cited in individual components is collected inside parentheses at the start of the assembly name.

Example (1)

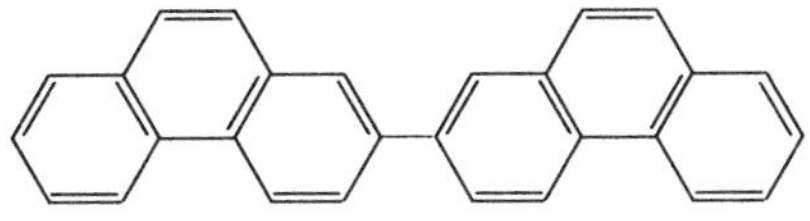

2,2'-Biphenanthrene

Example (2)

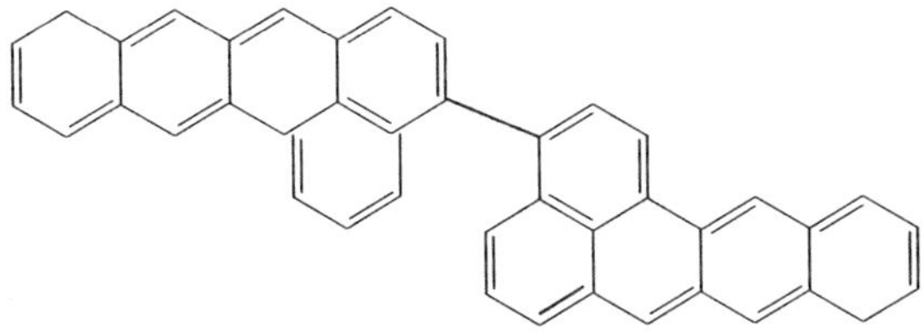

(9H,9'H)-3,4'-Bi(benzo[*de*]naphthacene)

Note that, even if the junction-locants of example (2) had been 9,9'-, those for the indicated hydrogen would still have been needed.

18.4 THE STRUCTURE HAVING NO DIRECT LINKAGE BETWEEN IDENTICAL UNITS

The name is based on the senior ring or ring-system, chosen according to the criteria of the yellow pages. All other ring-systems present are expressed as appropriately named radical prefixes (see Appendix C) and given the lowest available locant-set on the senior ring-system.

Example

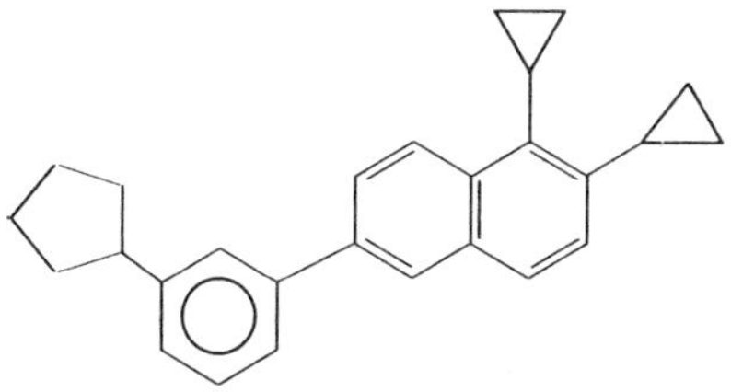

1,2-Dicyclopropyl-6-(3-cyclopentylphenyl)naphthalene
(1,2,6 preferred to 3,4,7 & 2,5,6)

19 Bridged systems

19.1 BICYCLO COMPOUNDS

These are characterized by being convertible into open-chain compounds by a minimum of two ring-scissions.

They are named by means of the word 'bicyclo', followed by a pair of square brackets containing three numerals separated by full stops [as for sections 13.3.1 and 13.3.2 (q.v.), except that the third numeral is not zero]. These represent the number of atoms in the bridges linking the two bridgeheads—the largest first and the shortest last. The name ending corresponds to the open-chain hydrocarbon having the same number of atoms as found in the cyclic structure. [See section 5.]

Double bonds are treated as for acyclic compounds; the ending becomes 'ene'.

Example (1)

Bicyclo[2.2.2]octane

Example (2)

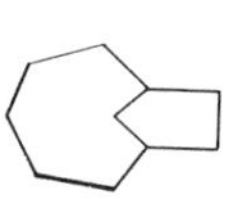

Bicyclo[5.2.1]decane

When heteroatoms occupy ring-positions, appropriate replacement prefixes from Table 3 are cited in the order given in Table 3, each with its appropriate locant. The name of the bicyclo all-carbon analogue, derived as above, then follows.

This also applies when the third numeral inside the square brackets [ ] is 0 in cases where one or more of the rings has fewer than five or more than ten members.

Example (3)

7-Oxabicyclo[2.2.1]heptane

Example (4)

5-Thia-1-azabicyclo[4.2.0]oct-2-ene

Numbering

This starts at one of the bridgeheads, follows the longest path to the other bridgehead, continues around the next longest bridge to the first bridgehead and then follows the shortest bridge. When there is a choice, lowest numbering is given to the following in the order listed:

  (i) heteroatoms taken as a set,
 (ii) the heteroatom earliest in Table 3,
(iii) unsaturation,
 (iv) substituents taken as a set,
  (v) substituents cited first alphabetically.

Example (5)

9,9-Dichlorobicyclo[6.3.2]dodeca-2,11-diene

[The bridgeheads are positions 1 and 7. If they were reversed, the double bonds would have the higher locants: 5,11.]

19.2 FUSED RING AROMATIC AND HETEROCYCLIC BASE-COMPONENTS

The mancunide †polycyclic systems from sections 12.1, 15, 16, 17.3, 18.1 and 18.2 may have bridging groups across any of their rings.

Typical hydrocarbon bridges are —CH_2— (methano); —CH_2CH_2— (ethano); —CH= CH— (etheno); —$CH_2CH_2CH_2$— (propano); —$CH_2CH_2CH_2CH_2$— (butano), etc.

Typical symmetrical hetero-bridges are —N=N— (azo); —O— (epoxy); —O—O— (epidioxy); —S— (epithio); —S—S— (epidithio); —NH— (epimino).

The positions of the two bridge-ends are denoted in the name by the lowest appropriate pair of locants taken from the unbridged parent structure consistent with fixed numbering constraints. [Follow the path through the flow-diagram at the back of the book as if the bridges were absent to obtain the appropriate unbridged name and numbering.] They are cited (separated by a comma) in ascending order if the bridge is symmetrical (if not, see the paragraph preceding example (3) of this section); followed by a hyphen and the name of the bridge (joined without a break to the name of the unbridged parent).

A comparison must be made between the bridged and unbridged structures. If the number of double bonds in the latter has been reduced in the bridged derivative by one, the name contains the prefix 'dihydro', if by two, 'tetrahydro', and so on. (Bridges between angular positions are rarely seen. If the structure has one, consult CNAS.)

Example (1)

9,10-Dihydro-9,10-ethanoanthracene

The 'dihydro', 'tetrahydro' etc., prefix is counted under 'h' among any other radical prefixes which may be present.

Different bridge-prefixes on the same base structure are cited in alphabetical order, each preceded by its locant-pair and a hyphen. Locant-pairs for repeated bridges are cited in ascending numerical order of the first numeral, each pair separated by a colon. In the event of a choice, the lowest set at the first point of difference is preferred.

Example (2)

1,2,3,4,10,10-Hexachloro-1,4,4a,5,6,7,8,8a-octachloro-1,4:5,8-dimethanonaphthalene

In example (2) the bridges are 1,4:5,8 whichever of the four possible sites is taken as

† Defined in section 1.5.

position 1. Final numbering is decided by the lowest locant-set for the substituent groups, and hydro prefixes all taken together.

When there is a choice, the locants for the bridges should be as low as possible. If more than one kind of bridge is present, lowest numbers are assigned to the bridgeheads in order of their citation in the name.

Numbering of the bridges: this continues from the highest number on the base-structure, proceeding from the higher to the lower numbered bridgehead.

Example (3)

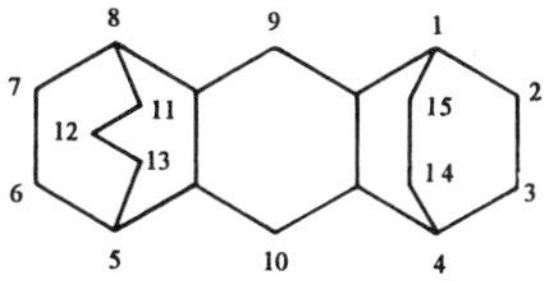

Perhydro-1,4-ethano-5,8-propanoanthracene

Non-symmetrical bridges may be unsaturated, e.g. `CH=CH—CH₂` (propeno), or compound bridges such as `O—CH₂CH₂` (epoxyethano), for which the name of the compound bridge is enclosed inside parentheses. For these the bridge-locants on the ring-system are cited in an order which corresponds to the senior end (e.g. the hetero-end) of the bridge coming first.

If in doubt as to the senior end of a non-symmetrical bridge, consult Table 3, and section B, yellow pages.

Example 4

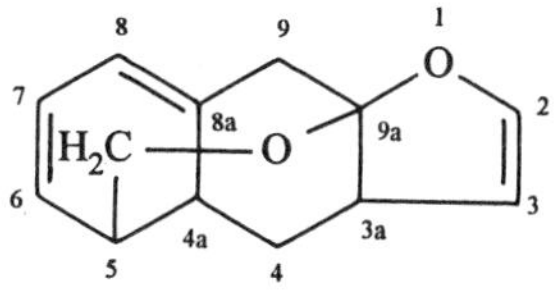

3a,5,9,9a-Tetrahydro-9a,5-(epoxymethano)naphtho[2,3-*b*]furan

19.3 CAGES

For more complex bridged systems, such as:

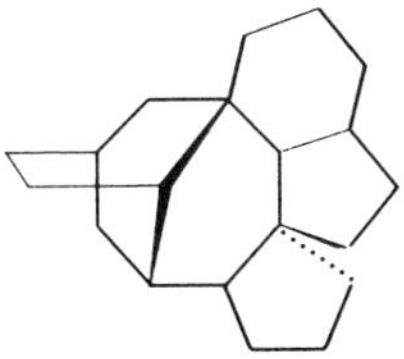

consult CNAS.

20 More complex polycyclic spiro-compounds

When one of the structures from sections 12.1,15,16 or 17.3 is linked to a second ring-system through a spiro-junction, the two components may be (a) identical before substitution or (b) different.

The site of spiro-attachment in the separated components may be –CH= or –CH$_2$–, as implied by the name, e.g. azulene or indoline, respectively. In the latter case the two H-atoms of the –CH$_2$– group are replaced by the spiro-junction but, in the former, the double bond has to be saturated or moved and the name reflects this modification either by uniqueness of the bonding in the spiro-compound or else by appropriate citation of indicated H in parentheses. When single, it is placed, preceded by its locant, immediately following the locant for the spiro-junction which generates it; when both systems are so affected, the Hs are cited in ascending order of their locants, separated by commas, then set inside parentheses after the spiro-locants.

20.1 SPIRO-LINKAGE BETWEEN IDENTICAL POLYCYCLIC COMPONENTS

The name is formed by citing (a) the spiro-locants separated by a comma and followed by a hyphen (b) "spirobi" and (c) the name of the ring-system (enclosed inside brackets if there is any possibility of confusion, e.g. from the conjunction of 'bi' and 'cyclo').

Example (1)

1,1′-Spirobi-indene

In example (1) the name "indene' implies a $-CH_2-$ site at position 1 but this satisfies the spiro-formation in both components. The bonding is thereby fixed and so the full name needs no H-indicators.

Example (2)

2,2'-Spirobi[cyclopentacyclo-octene]

In Ex. (2) the name "cyclopentacyclo-octene" implies a $-CH_2-$ position but, as for example (1), this coincides with the spiro-site and needs no citation.

Example (3)

1,2'(8'*H*,8a*H*)-Spirobinaphthalene

In Example (3) the name "naphthalene' implies a $-CH-$ site at the spiro-junction. The bond-rearrangement now accommodating it has involved generation of an additional $=CH_2-$ site in each system and these are indicated in parentheses immediately following the spiro-locants, as shown.

This example illustrates the choice of the lowest possible locants for the spiro-junction (1 rather than e.g. 4 and 2' rather than 3'). In this case the naphthalene fixed numbering leaves no choice for the indicated H locants but, if a choice is presented, they are chosen to be the minimum consistent with the fixed numbering constraints.

20.2 SPIRO-LINKAGE BETWEEN DIFFERENT POLYCYLIC COMPONENTS

The name is here formed by citing in order (a) "spiro[" (b) the name of the component earliest in alphabetical order* (c) the spiro attachment-locants of each component, separated by a comma and flanked by hyphens (d) the name of the second component (e) "]". The spiro-locants are chosen to be as low as fixed numbering constraints allow, the first unprimed, the second primed.

(* For this procedure to be internationally meaningful the English alphabet and English spelling have to be used for naming these compounds.)

Example (4)

Spiro[1,3-dioxolan-2,3'-indole]

In example (4) order of citation is 'd' before 'i'. Although the name 'indole' implies an H-atom on position 1, the spiro-formation specifies the tautomer shown; no H-terms are needed.

20.3 SUBSTITUTED AND HYDROGENATED DERIVATIVES

The skeletal structure of the spiro-compound is treated as integral (a) for naming substituted derivatives and (b) in comparison with other ring-systems in the same structure for priority considerations (as in Section A, Yellow Pages).

However, names formed according to section 20.1 are usable only if their hydrogenation-state is identical or if any differences are without effect on the name-stem or its ending. If dihydro, tetrahydro, etc. prefixes have to be used, they are cited (preceded by appropriate locant-sets in ascending order and followed by a hyphen) in alphabetical order along with any other prefixes for attached substituents. Locants for the system named second are primed and a primed locant is cited after the same numeral unprimed but before higher ones.

Example (5)

5'-Bromo-1,1',2',9a-tetrahydro-8-methyl-2,7'-spirobi(benzocycloheptene)
(prefix order:b,h,m. The structure is that of a tetrahydro-derivative of 2,7'-spirobi(benzocycloheptene),
which has no –CH$_2$–positions.

If different hydrogenation-states affect the name-stem or its ending, the procedure of section 20.2 is followed.

Example (6)

7',7a'-Dihydrospiro[indan-1,1'-indene]

In example (6) "indan" is appropriate. Similar considerations apply to the pairs indole/indoline, isoindole/isoindoline, chroman/chromene and isochroman/isochromene.

Example (7)

2′,3′,4′,5′-Tetrahydrospiro[benzo[*b*]furan-2(3*H*),3′–furan]
(citation order: benzofuran before furan.)

Note that, in Example (7) the 'tetrahydro' is cited at the start and not as part of the spiro-system.

Example (8)

3′,6-Dichloro-4′-(4-chloroacridin-9-yl)spiro[morpholine-3,2′(3′*H*)-quinoline]
(prefix order: chloro, chloroac)

In example (8) the name "morpholine" implies a $-CH_2-$ site at its position 3 whereas "quinoline" implies a $=CH-$ site at its position 2. Spiro-formation has, in this instance, generated a $-CH_2-$ site at 3, which is cited as shown even though this H-atom has been substituted. Citation order inside the [ ] is 'm'before 'q'. The spiro-skeleton is deemed senior to acridine by criterion A(8), Yellow Pages.

Derived radicals: These are named by adding the locant for the free valence flanked by hyphens, then "yl" after the full name of the spiro-compound. If this ends with 'e', it is elided before the 'yl'.

Example (9)

4-(Spiro[morpholine-3,2′(4a′*H*)-naphthalen]-4′-yl)phenoxazine

In example (9) the senior structure is decided by criterion (14) of section A, Yellow Pages.

Seniority of ring-systems, chains and ring-with-chain systems

A. RING-SYSTEMS

In deciding the seniority of ring-systems, whether identifying the senior component in section 17.3 or determining the basis of the name for a molecule having separated diverse ring-systems, the reader should apply the following criteria, as applicable, *in the order given* until a decision is reached.

(1) The ring or ring-system having the greatest number of PGs collectable as name-endings (see 'collected' in section 1.5) is senior.

Example

(2) The trivially named vari-functional structures named in section 23 are preferred over homofunctional structures with an equal number of PGs (but see section C).

(3) Heterocycles, however simple, are senior to all carbocycles, however complex.

For heterocycles:

(4) Rings and ring-systems having one or more nitrogen atoms in a ring position are senior to those with none.

Example

(5) In the absence of nitrogen atoms in ring positions, those with a heteroatom as high as possible in Table 3 are senior.

Example

Thereafter, the senior component is that with:

(6) The greatest number of rings.
(7) The largest possible individual ring.
(8) The greatest number of ring-heteroatoms.
(9) The greatest variety of heteroatoms.

Example
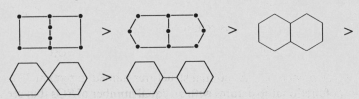
is senior to

(10) The greatest number of ring heteroatoms listed earliest in Table 3.
(11) The lowest locant-set for heteroatoms before fusion (this criterion applies only to choice of senior component for fusion).

For carbocycles the named structures (1)–(12) of section 18.1 are given in ascending order of seniority, applied regardless of whether the following criteria give the same result or not.
For other carbocycles the senior component is that having:

(12) The greatest number of rings.
(13) The largest individual ring at the first point of difference.

For carbocyclic and/or heterocyclic systems the component having:

(14) The largest number of atoms in common among rings

Examples in descending order of seniority:

(15) The earliest letters in the fusion-locants [a<b].
(16) The lowest numbers in the expression for the ring-junction, whether fusion, spiro-, or bicyclo-, etc.
(17) The lowest state of hydrogenation.
(18) The lowest locant for indicated hydrogen.
(19) The lowest locant(s) for directly attached PG(s).
(20) The greatest number of attached groups expressed as prefixes.
(21) The lowest locant-set for prefixes, dihydro-, etc., groups and unsaturation-locants, all taken together.
(22) The lowest locant for the prefix cited first in the name.

Note
Assemblies of benzene rings are treated for seniority as a single system consisting of two rings ('biphenyl') or three rings ('terphenyl') etc. In a spiro-compound any rings fused directly to either of the rings involved in spiro-union are included for consideration along with them both under this set of criteria.

For others the repeated parent structure ('PS') [see Appendix D] is considered for seniority assessment as if it were alone. Only when the same PS occurs both alone and re-duplicated as an assembly in the same molecule is the assembly considered as such for seniority purposes.

B. CHAINS

Many of the functional groups of Table 1 (final column) can be attached to any chain-atom, including end-positions (e.g. $-OH$, $-SO_2OH$, $-NH_2$).

Some can be considered as occupying the end-position (e.g. $-COOH$, $-CN$, $-C(=NH)NH_2$, and some can appear in mid-chain (e.g. $-C(=O)-$, $-NH-$, $-C(N=OH)-$). Certain combinations of heteroatoms in adjacent positions in a chain constitute groups of this third class (for example, $-O-O-$, peroxide; $-NH-NH-$, hydrazine) and Table 1 should be carefully examined at the Box (iv) stage when the **flow-diagram** is being explored.

Subject to these considerations, for the purposes of these criteria, a carbon-atom chain terminates at the first heteroatom encountered. However, if a carbon-atom chain contains more than three heteroatoms in non-terminal positions, the entire chain is considered for seniority and named (by replacement nomenclature, see section 4) as if it were an all-carbon chain. Functional endings, e.g. 'amine', 'ether' and 'sulphide' are not then used.

Example (1)

$$CH_3-O-CH_2-S$$
$$CH-SiH_2-CH_2-CH_3$$
$$CH_3-CH_2-O$$

No one chain has more than three such heteroatoms, and the name is:

ethyl (ethylsilyl)(methoxymethylthio)methyl ether

ethyl being preferred to methyl, so that the name is based on the lower of the two possible ethers, as depicted.

Example (2)

$$CH_3-O-CH_2-S$$
$$CH-SiH_2-CH_2-CH_3$$
$$CH_3-NH-CH_2-O$$

The left-hand side constitutes a chain containing four non-terminal heteroatoms and the name is, accordingly:

5-(Ethylsilyl)-2,6-dioxa-4-thia-8-azanonane.

The grouping $-\overset{\|}{\underset{O}{C}}-O-(C)$ denotes an ester and the chain in which it occurs should be split between the $\overset{O}{\overset{\|}{C}}$ and the $-O-$ and the two fragments assessed separately for the 4-heteroatom criterion.

Subject to the above special provisions, the following criteria are applied **in order** until a decision is reached.

(1) The chain carrying the greatest number of directly attached PGs (see Introduc-

tion, section 1.5) [1 is more than 0].

(2) The trivially named vari-functionals of sections 22 and 26 are preferred to homo-functional structures with an equal number of PGs, but see section C.

(3) The chain containing the greatest number of non-terminal heteroatoms (when there are four or more).

(4) The chain containing the greatest number of heteroatoms occurring first in the order of Table 3.

(5) The chain having lowest locants for the heteroatom set.

(6) The chain having the lowest locant(s) for the heteroatom(s) cited earliest in criterion (4).

(7) The chain containing the maximum number of unsaturation-sites, whether double or triple bonds.

(8) The longest chain of carbon atoms. [If a carbon-chain contains more than three heteroatoms in non-terminal positions, it is counted for this criterion as if it were an all-carbon chain—but read the Introduction to section 4 before proceeding.]

(9) The chain with the greatest number of double bonds.

(10) The lowest locant-set at the first point of difference for the PGs.

(11) The chain having the lowest locant-set, at the first point of difference, for its multiple bonds.

(12) The chain having the lowest locant-set, at the first point of difference, for its double bonds.

(13) The chain with the most substituents cited as prefixes.

(14) The chain having the lowest locant-set for all the prefixes at the first point of difference.

(15) The chain bearing the substituent cited earliest in alphabetical order.

(16) The chain bearing the lowest locant for the substituent cited earliest in alphabetical order.

C. SENIORITY OF RING-SYSTEM vs. CHAIN [cf. section 10]

For the purposes of this section, the most senior ring system and the most senior chain are each decided first according to sections A and B respectively and *then* compared.

(1) The ring-system (whether monocyclic or polycyclic) or chain bearing the maximum number of collectable PGs is senior.

(2) The trivially named multi-functionals of sections 22, 23 and structures (1), (2) and (8) of section 26.1.1 are preferred to homo-functional structures with an equal number of the same PG.

Between them, seniority of acids is as follows:

   (i) glutamic acid
   (ii) aspartic acid
   (iii) tartaric acid
   (iv) dicarboxylic acids following criteria (C-3)–(C-6), then (A-3)–(A-22) or (B-3)–(B-16), as appropriate
   (v) acids named in sections 22.2.1 and then those in 22.2.2
   (vi) acids of section 22.1
   (vii) other acids of section 22 or those of section 23.

Within each of categories (v)–(vi) the A or B criteria-set is applied, as appropriate, until a decision is reached.

Seniority of ring-system vs. chain

To resolve choices in category (vii) the C-criteria have to be applied first

(3) If the PG-number is equal (or zero for each), the system with the greatest number of skeletal atoms in the ring-structure or the unsubstituted chain is senior.
[*Exceptions*. If there are no FGs and a chain has the same ring-system in the same hydrogenation-state multiply attached to it, the chain is senior.] Equally, if a ring-system has the identical chain multiply linked, the ring-system is senior. Examples:

2-(4-quinolyl)-5-(quinolyl)heptane;

1,3-bis(2-chlorodec-6-enyl)benzene.

(for these exceptions to apply to substituted cases, the pattern of substitution must also be identical.)

(4) The ring or ring-system or chain carrying the greatest number of substituents expressed as prefixes is senior.

(5) The ring, ring-system or chain having the greatest variety of prefixes is senior.

(6) The ring-system is senior.

21 Radicofunctionals

21.0 GENERAL NOTES

Radicofunctional names are constructed from appropriate names for radicals and functional class-names. The latter are given in the following sections (see sections 21.1 to 21.17). Names for radicals are to be found in Appendix C, in Table 2 and in the third column of Table 1.

Where more than one radical-name is cited, the order is alphabetical.

If the PG occurs more than once in the structure and the method of section 4 (q.v.) is not appropriate, the most central PG is chosen for citation as the name-ending in the cases appropriate to sections 21.3, 21.4, 21.5, 21.11, 21.12, 21.13 and 21.14.

21.1 CATIONS

R^1–R^4 are monovalent organic radicals but, in addition to being in this environment, the charged nitrogen atom may also occupy a ring position, in which case R may be H.

In the first instance, the entity is in effect an ammonium ion $(NH_4)^+$, whose hydrogen atoms have been replaced by organic groups.

The cation is named by citing the names of the attached radicals in alphabetical order before 'ammonium'. Ambiguity is avoided by enclosing later-cited groups in parentheses (unless they are preceded by di or tri).

Cations are found in association with anions and the name is completed by citing the anion after the cation, separated by a space.

Example (1)

Benzyltrimethylammonium bromide

Example (2)

Bis[allyl(chloromethyl)(cyclobutyl)(methyl)ammonium] sulphate

When the cation is based on aniline or its analogues (see section 21.9.1) or where the N^+ ion occupies a ring-position the name of the cation is formed from the name of the uncharged free base by changing its 'ine' ending to 'inium'. Names for any substituent groups, whether on the charged site or elsewhere, are expressed as radicals (see Appendix C) in alphabetical order.

Example (3)†

4-Iodopyridinium iodide

Example (4)

1-Ethyl-1,4-dimethylpiperidinium chloride

Example (5)

N,N-Dipropyl-*p*-toluidinium chloride

---

† Because cations are widely separated from amines in Table 1, this method of naming has the effect—when an FG senior to amine is also present—of drastically altering the name of a base when its salts are formed. It may be preferable, e.g. for indexing and listing, when a cation is formed by proton acquisition, to retain the name of the neutral base and add, after a space, the name of the acid moiety. Thus structure (3) could be 4-iodopyridine hydroiodide and structure (5) could be *N,N*-dipropyl-*p*-toluidine hydrochloride.

21.2 ACYL HALIDES

These have the general formula: $R—C(=O)—X$, where X is F,Cl,Br or I. Their names are radicofunctional in form and consist of the name for the appropriate acyl radical, a space, and then the halide name. They are treated in more detail in sections 22.4.2, 23.1.5.3, 24.10 or 25.10, as appropriate.

21.3 RING-FLANKED KETONES AND THIOKETONES

Where one side of the $>C=O$ is an open chain see section 24.18.2. Where both sides are open chains see section 25.18.

21.3.1 Unsubstituted monoketones

(For polyketones see section 27.4 if collectable* or section 21.3.3 if not collectable.) In the special circumstances where two separate ring-systems are linked directly by a $>C=O$ group (each at a carbon atom site) the name is formed of three words. The first two are the two ring-systems separately named in the form of the appropriate radicals (see Appendix C), cited in alphabetical order. The third word is 'ketone'; for $>C=S$ it is 'thioketone'.

If either ring-system is joined to the $>C=O$ group at a heteroatom-site, turn to section 24.18.1(b).

Example (1)

Cyclopent-1-enyl 2-pyridyl ketone

In the event that the two ring-systems are identical† and also joined to the central –CO– group in the same way, they can be collected by means of the prefix 'di-'.

Example (2)

Di-3-furyl ketone

Example (3)

2-Naphthyl phenyl thioketone

*'Collected' is defined in section 1.5.
†This includes identically substituted, in which case 'bis' replaces 'di'.

Exception (1)

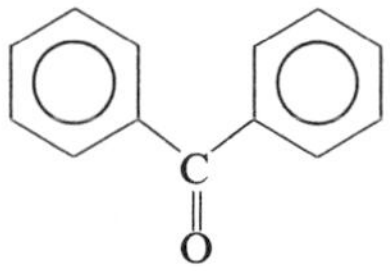

Benzophenone

Exception (2)

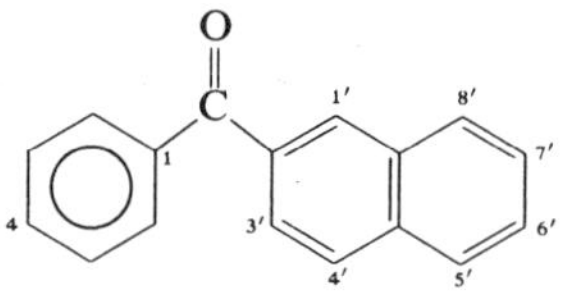

Benzo-2′-naphthone
[Also (3) benzo-1′-naphthone]

The $>$C$=$S analogues of these structures are named diphenyl thioketone and phenyl 2-(or 1-)naphthyl thioketone, respectively.)

21.3.2 Substituted derivatives

See Appendix C-0

Provided that the substituents on the two-ring systems of section 21.3.1 do not constitute or contain one or more functional groups 'senior' to ketone in Table 1, they are expressed in the name in prefix form. Example (4) in this section illustrates the use of 'bis' for cases where the pattern of substitution and mode of linkage to the central (CO) are identical for both rings. When one of these conditions is not fulfilled, the three-word method has to be used and the alphabetical order of citation will be decided by the first letter of each cyclic radical, complete with its named substituents. If they start in the same way, then comparison is continued until a difference is reached and that decides the order.

Example (1)

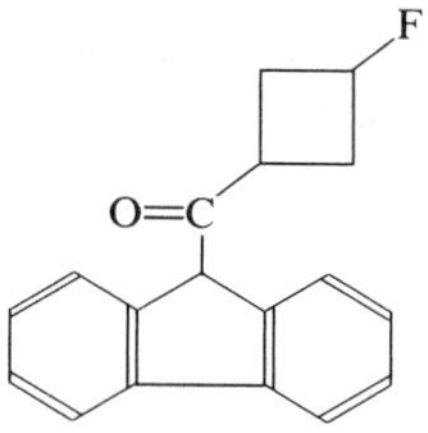

9*H*-Fluoren-9-yl 3-fluorocyclobutyl ketone
(the sixth letter decides the order: 'e' before 'o')

Example (2)

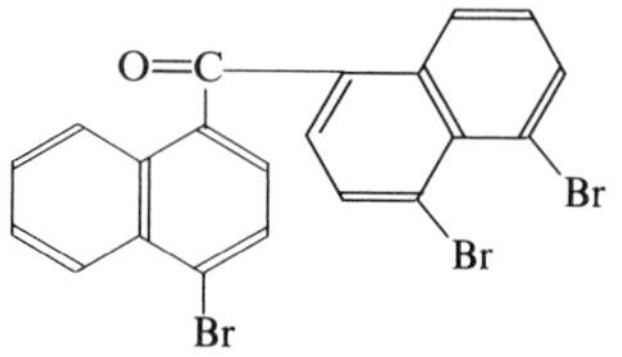

4-Bromo-1-naphthyl 4,5-dibromo-1-naphthyl ketone
('b' before 'd')

If the only difference between the two cyclic systems is in their position of linkage, then the order is 'lowest locant first', e.g.

Example (3)

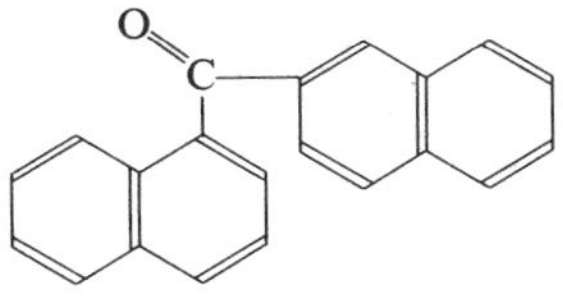

1-Naphthyl 2-naphthyl ketone
(1 before 2)

Example (4)

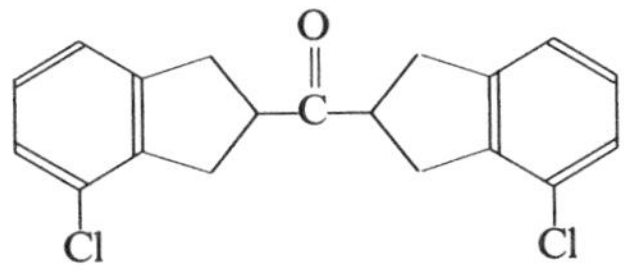

Bis(4-chloroindan-2-yl) ketone

Substituted benzophenones are named by citing all substituents in alphabetical order (assuming no groups senior to ketone are present), and each is assigned its number as follows:

the ring with the fewest substituents has its locants (numbers) designated by primes; that bearing the greatest number has 'unprimed' locants. Numbering then proceeds around the ring, starting at the position joined to the central (CO) group and following the direction which gives the lowest number to the first group encountered. In symmetrical cases it will be the same whichever direction is chosen.

Example (5)

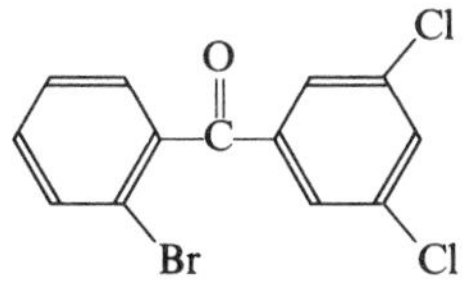

2'-Bromo-3,5-dichlorobenzophenone
(The Cl atoms are in positions 3 and 5 regardless of the direction but, for the Br atom, 2' is less than 6'.)

In cases where the set of ring-locants is the same for each ring, that which is cited first according to alphabetical order has the unprimed numbers. Thus, in the following example:

Example (6)

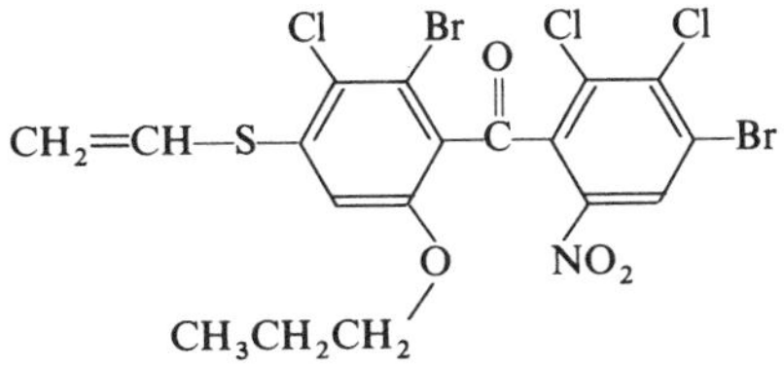

the lowest numbers obtainable for the sets of substituents on each ring is 2,3,4,6- (in both cases the alternative direction gives 2,4,5,6, which is 'higher' at the first point of difference and so not preferred.) However, examination of each ring in turn shows that 'bromo' will be the earliest cited (alphabetically), and so the ring with the 2-bromo will be 'unprimed' and the other will be 'primed' because 4-bromo is later. Once this is decided, the groups on both rings are cited alphabetically, collecting 'like' groups by the prefixes di, or tri etc., and using ascending order for the appropriate numerals. For this purpose, 'primed' numbers are taken as 'higher' than the same numbered 'unprimed', but lower than the next numeral up, whether 'primed' or not.

Example (6) is thus named 2,4'-Dibromo-2',3,3'-trichloro-6'-nitro-6-propoxy-4-(vinylthio)benzophenone. (The vinylthio is enclosed inside parentheses (a) because this is the composite group on position 4 and (b) to avoid possible ambiguity—it could otherwise be taken that the 'thio' implied that the ketonic oxygen had been replaced by sulphur.)

Other ketones are not named according to the above-described methods but are dealt with under their appropriate sub-sections, namely 24.18, 25.18, 26.2 and 27.4.

21.3.3   Polyketones, non-collectable *

If the structure contains a $>$C=O group in a mid-chain position, the name is formed according to the procedure of section 25.18 and the cyclic ketone fragments are named along with any other substituents, cited as attached groups in alphabetical order.

Example (1)

4-[3-(Cyclopentylcarbonyl)cyclohexyl]butan-2-one

If a $>$C=O group occupies a ring-position, the name is formed according to the procedures of section 24.18.3. If mid-chain and ring-site $>$C=O groups are both present, the name is decided by application of the C-criteria of the Yellow Pages. For this the $>$C=O flanked by two ring-systems is counted as a chain of one member.

Example (2)

3-Acetyl-4-benzoylcyclopenta-2,4-dienone
[senior structure decided by C(3)]

* "Collected" is defined in section 1.5.

If the only carbonyl groups present are ring-flanked or at a chain-end, the name is based on the most central, using the procedure of section 21.3.2, the remaining carbonyl groups being cited in prefix-form, making use of allowed contractions (such as benzoyl) where appropriate.

If there is a choice of central carbonyl groups, the names formed using the method of section 21.3.1 are compared and that coming earliest in an alphabetic index is preferred.

The '-phenone' and '-naphthone' names of 21.3.1 and 24.18.2 are used when appropriate provided this does not conflict with the instructions of this section (except the decision based on alphabetical order, which is used as a last resort).

Example (3)

2-Hydroxy-3'-(2-furoyl)benzophenone

Example (4)

4-Benzoylcyclopent-2-enyl 5-acetyl-6-methyl-3-piperidyl ketone

Example (5)

4-(4-Benzoylbenzoyl)tetrahydro-3-furyl 3-benzoylcyclopentyl ketone   (i)
4-(3-Benzoylcyclopentylcarbonyl)tetrahydro-3-furyl 4-benzoylphenyl ketone   (ii)
[(i) preferred: benzoylb, benzoylc]

The names of the $>C=S$ analogues of these examples are as follows:

Ex.(1)   4-{3-[Cyclopentyl)thiocarbonyl]cyclohexyl} butane-2-thione
Ex.(2)   3-Thioacetyl-4-thiobenzoylcyclopenta-2,4-dienethione
Ex.(3)   2-Furyl 3-[(2-hydroxyphenyl)thiocarbonyl]phenyl thioketone
Ex.(4)   6-methyl-5-thioacetyl-3-piperidyl 4-[(phenyl)thiocarbonyl]cyclopent-2-enyl
          thioketone

Ex.(5)   3-[(phenyl)thiocarbonyl]cyclopentyl 4-(tetrahydro-4-{4-
         [(phenyl)thiocarbonyl]phenyl}thiocarbonyl)-3-furyl thioketone.

[(3) and (5) are decided by the alphabetical preference. Note also the use of enclosing marks to avoid ambiguity. The group C_6H_{11}–S–C(=O)– is named "(-cyclo-hexylthio)carbonyl".]

These are of the general formula R^1—C(=S)—R^2 (although R^1 and R^2 may be the same).

Names are formed exactly as under sections 21.3.1 and 21.3.2 except that the final word 'ketone' is replaced by 'thioketone'.

21.4   HYDRAZONES

These are of the general formula:

$$\begin{array}{c} R \\ \diagdown \\ \diagup \\ R \end{array} C{=}N{-}NH_2$$

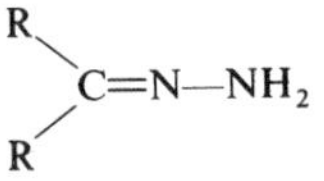

but the R-groups may be the same or different and the hydrogen atoms may be substituted by other groups.

The name is formed by first replacing everything to the right of the double bond by an oxygen atom and then naming the resultant ketone—after progress through the **flow-diagram** at the back of the book to the relevant section. The name so derived is followed by a space and then 'hydrazone' in the case of the structure shown above or, if the nitrogen atom bears substituents, their radical names (in alphabetical order if diverse) and then 'hydrazone' without a break.

Example (1)

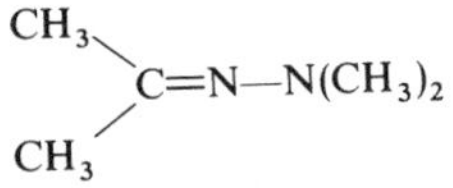

$$\begin{array}{c} CH_3 \\ \diagdown \\ \diagup \\ CH_3 \end{array} C{=}N{-}N(CH_3)_2$$

Acetone dimethylhydrazone
(Name derived from a ketone of section 25.18)

Example (2)

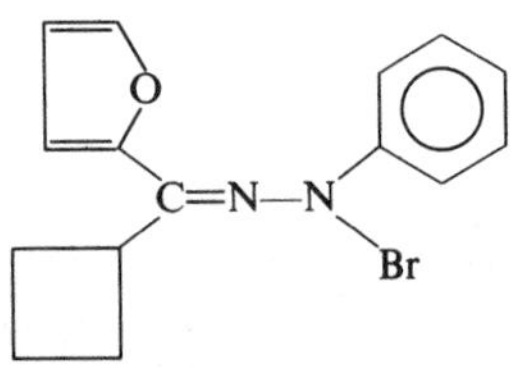

Cyclobutyl 2-furyl **ketone** bromo(phenyl)hydrazone
(Name derived from a ketone of section 21.3.1)

These may be aldoximes of general formula R—CH=N—OH, or ketoximes:

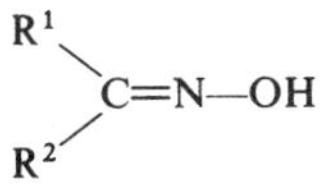

$$R^1 \diagdown\!\!\!\diagup R^2 \; C=N-OH$$

21.5.1 Unsubstituted oximes

These are named by adding, after a space, the word 'oxime' to the name of the corresponding aldehyde (R—CH=O) or ketone:

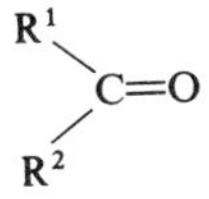

$$R^1 \diagdown\!\!\!\diagup R^2 \; C=O$$

which is first formed according to the appropriate rule section (mono-PG rules; aldehydes, or ketones, see sections 21.3, 22.4.5, 23.1.3, 24.17, 24.18, 25.17, 25.18).

Example (1)

$$CH_3—CH_2—CH=N—OH$$

Propionaldehyde oxime

Example (2)

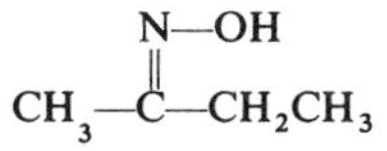

$$\overset{\displaystyle N—OH}{\underset{}{CH_3—\overset{\|}{C}—CH_2CH_3}}$$

Butanone oxime
(Name derived from a ketone of section 21.3)

Example (3)

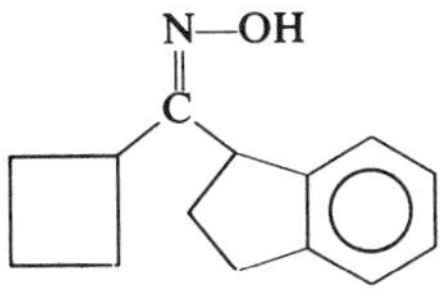

Cyclobutyl indan-1-yl ketone oxime

21.5.2 Substituted derivatives of oximes

In addition to substitution on carbon atoms, the hydrogen atom of the –OH group may also be replaced by a group or atom. When this is the case, the name is formed by citing the name of the corresponding aldehyde or ketone, followed by a space, the locant *O*- and then, immediately after the hyphen, the name of the O-attached radical; and finally, without a break, 'oxime'

Example (1)

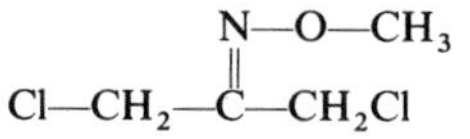

1,3-Dichloroacetone *O*-methyloxime

Example (2)

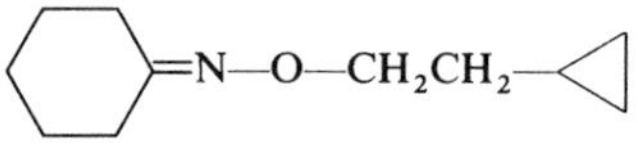

Cyclohexanone *O*-(2-cyclopropylethyl)oxime

Example (3)

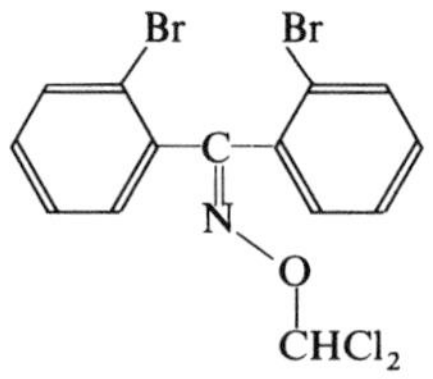

2,2′-Dibromobenzophenone *O*-dichloromethyloxime

Example (4)

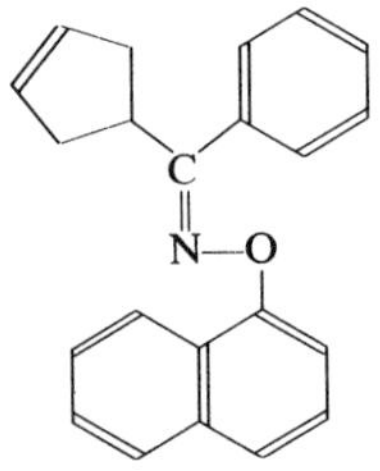

Cyclopent-3-enyl phenyl ketone *O*-(1-naphthyl)oxime

21.5.3 Stereoisomers

When R^1 and R^2 in the general formula for oximes are not identical (i.e. unlike examples (1)(2) & (3) of 21.5.2), the –OH or –OR group may be aligned towards the more senior group or the less senior. The senior group is decided by means of the bond-exploration procedure described under section 2.2, q.v., if not immediately apparent. Alignment towards the senior group is denoted by (*Z*)- at the start of the name; the alternative by (*E*)-. Thus examples (2) and (3) of 21.5.1 are shown as (*Z*), as is example (4) of 21.5.2, whilst Ex.(1) of section 21.5.4 illustrates (*E*)-alignment also.

21.5.4   Polyoximes

If the structure has more than one oxime-grouping and if, after formal replacement of each
=N–OH or =N–OR group by=O, the resulting polyketone can be given a name collecting
them, the name of that polyketone is followed by a space and then "dioxime", "trioxime",
etc. as appropriate.

(This assumes that there are no 'extra' =O groups in the original structure. In that case,
oxime is not the PG, the name will be based on the ketone and the oxime-groupings expressed
in it as hydroxyimino- or R-yloxyimino-prefixes.)

Example (1)

$$\begin{array}{cc} \text{N} - \text{OH} & \text{N} - \text{OH} \\ || & || \\ \text{CH}_3 - \text{C} - \text{CH}_2 - \text{CH}_2 - \text{C} - \text{CH}_3 \end{array}$$

(*Z,E*)-Hexane-2,5-dione dioxime

Where the oxime-sites are interchangeable due to symmetry [as in ex.(1)] *Z* is cited first, as
shown. Otherwise, the stereodesignators are cited in order of their locants and cited
immediately after them (see example 2). In cases of O-substitution, the locants *O* and *O'* are
used, the prime having the significance of the higher-numbered locant.

Example (2)

$$\begin{array}{cc} \text{OE}_t & \text{OBu} \\ \diagup & \diagup \\ \text{N} & \text{N} \\ || & || \\ \text{CH}_3\text{CH}_2 - \text{C} - \text{CH}_2\text{CH}_2\text{CH}_2 - \text{C} - \text{CH}_3 \end{array}$$

(2*E*, 6*Z*)–*O*–Butyl–*O'*-ethyloctane-2,6–dione dioxime

21.6   RADICO-ALCOHOLS

The only alcohols to be named under this section are:

(1) Benzyl alcohol

(2) Benzhydryl alcohol

(3) Phenethyl alcohol

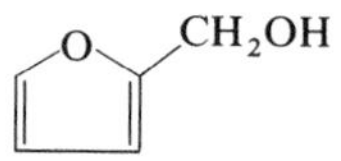

(5) Furfuryl alcohol 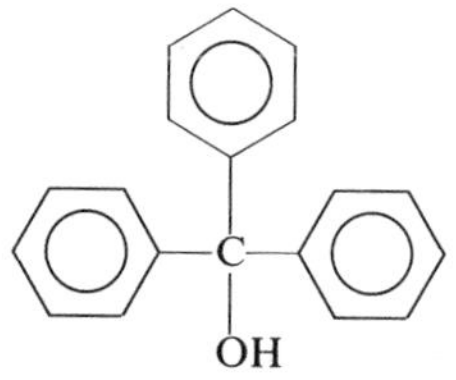   (4) Trityl alcohol

In the event of a choice, (5) is preferred to (4), (4) preferred to (3), etc., but first read the provisos at the end of this section.

Substituted derivatives of these alcohols are named according to Appendix C-0, (q.v.). The only space in the name is that before 'alcohol'. When there is a choice of numbers around a ring, the lower is chosen, (i.e. 2 is lower than 6, and 3 is lower than 5).

If there is a symmetrical distribution of substituents around the ring (e.g. 2 and 6; 2,4 and 6; 2,3,5 and 6), the numbering is chosen which will be the lower for the substituent first cited in alphabetical order in the name.

Example (6)

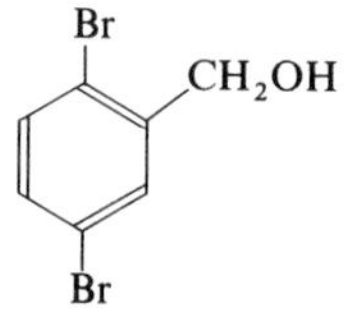

2,5-Dibromobenzyl alcohol
(and not 3,6-)

Example (7)

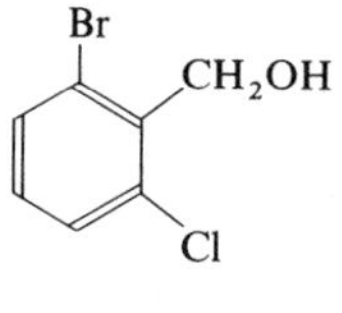

2-Bromo-6-chlorobenzyl alcohol
(and not 6-bromo-2-chloro ...)

Example (8)

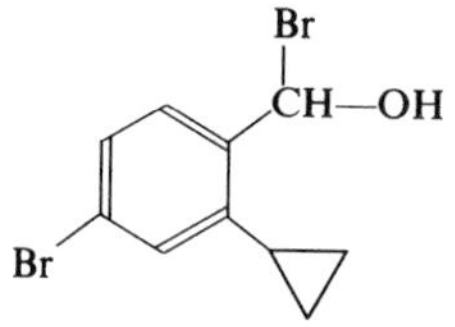

α,4-Dibromo-2-cyclopropylbenzyl alcohol
(α,2,4 preferred to α,4,6)

Example (9)

4,4′-Dibromo-3-(3,3-difluoropropyl-α,3′-difluoro-5-nitrobenzhydryl alcohol
(prefix order: b,d,f,n)

Example (10)

α-Phenylfurfuryl alcohol

In the case of trityl alcohol, the locants for the three rings are 'unprimed', 'primed' and 'double-primed', respectively, according to the pattern of substitution. Criterion (A-21), yellow pages requires that 'primed' is 'higher' than 'unprimed' (see also section 23.2).

These instructions are subject to two provisos:

(a) No group senior to alcohol in Table 1 is present.
(b) No carbon atoms are attached directly to the α-carbon atom, in each case, unless such atoms occupy ring-positions in a cyclic structure.

Thus, benzyl alcohol would still be the name-ending if the side-chain Br of Example (8) were replaced by a cyclohexyl group for instance, but not if it were replaced instead by a methyl group, when ethanol would then become the name-ending. If it were replaced instead by a phenyl or a tolyl group, the name would then end in 'benzhydryl alcohol'. Similarly, if the lone side-chain fluorine atom of Example (9) were to be replaced by an ethyl group, the compound name would have to be based on propan-1-ol, (see section 25.20.1). Again, any cyclic radical (except phenyl or 2-furyl) could replace this fluorine atom so long as it is joined directly from a ring-position and not via a side-chain. A further exception is that, if any benzene ring of structures (1)–(4) has another benzene ring directly attached to it, that pair of rings is named 'biphenyl-*x*-yl' where '*x*' denotes the locant of the attachment-position. In such cases, names for (1,2 and 4) are based on methanol and names for (3) based on ethanol (see section 25.20.1).

21.7   ESTERS OF INORGANIC ACIDS

Such acids may be monobasic (e.g. nitric acid), dibasic (e.g. sulphuric acid) or of higher basicity (e.g. phosphoric acid). The name is formed by citing in order: (a) the numerical prefix for the number of radicals in the ester ('mono' is always omitted); (b) the name of the radical (see Appendix C); (c) a space; (d) 'hydrogen' if it is an acid ester (preceded by di, tri, etc. as appropriate)—in which case, another space; and (e) the name of the appropriate anion.

Example (1)

$$C_6H_5CH_2CH_2{-}ONO_2$$

Phenethyl nitrate

Example (2)

$$C_6H_5CH_2{-}O{-}S(OH){=}O$$

Benzyl hydrogen sulphite

Example (3)

$$(C_6H_5)_2CH{-}OP(OH)_2{=}O$$

Benzhydryl dihydrogen phosphate

Example (4)

$$\begin{array}{c} C_6H_5CH_2O \\ \diagdown \\ \qquad\qquad C{=}O \\ \diagup \\ C_6H_5CH_2O \end{array}$$

Dibenzyl carbonate

The anionic moiety may be represented in various ways, e.g.

$$-OSO_2H, \quad -(HSO_3), \quad -O{-}S(OH){=}O, \quad -O{-}\overset{\uparrow}{\underset{}{S}}{-}OH \quad or \quad -O{-}\overset{}{\underset{\downarrow}{S}}{-}OH$$

Confusion is best avoided by redrawing linear forms in the more extended manner showing all relevant connectivities before attempting the identification.

21.8 HYDROPEROXIDES

If this is the PG or the only functional group present capable of being used as a name-ending, the name is formed by citing the name of the attached radical (see rules for radicals) then, following a space, the word 'hydroperoxide'.

Example

3-Chloroquinolin-5-yl hydroperoxide

21.9 AMINES

$$RNH_2; \quad R^1R^2NH; \quad R^1R^2R^3N$$

The three types of amine indicated by these generalized formulae are known as primary, secondary and tertiary amines, respectively. The R-symbols stand for monovalent organic radicals and, in the last two classes, they may be the same or different. They may be considered as being derived by substitution into the parent hydride, NH_3, which is called **ammonia** when unsubstituted.

However, in its organic derivatives, this is shortened to 'amine', so that CH_3—NH_2 is known not as methylammonia but as methylamine. This style of naming may be regarded as a kind of radicofunctional one but with the peculiarity that there is no space in the name.

Note that cyclic amines such as structures (19) and (20) of section 9.1 are regarded, when using the **flow-diagram** at the back of the book, as having no FGs *per se*.

21.9.1 Trivially named phenyl monoamines

The following structures have trivial names retained as parents for derivatives, whether substituted on the nitrogen atom or the ring

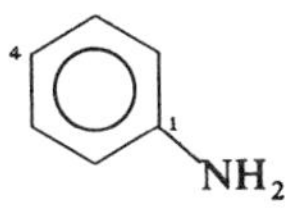

(1) Aniline

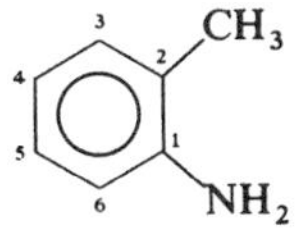

(2) Toluidine
[*o*- shown]

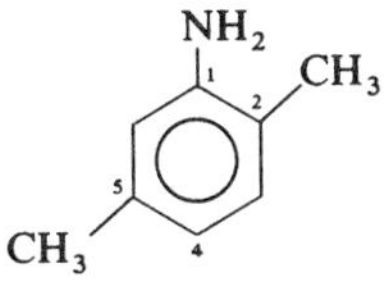

(3) Xylidine
[2,5- shown]

These names are preferred as parents when appropriate and, in the event of a choice, (3) is preferred to (2) and (2) to (1) as PFS.† Between isomers, that having the lowest numbering is preferred (*m*- is 'lower' than *p*- and *o* is 'lower' than *m*).

The NH_2 group takes position 1 and the CH_3 groups have numbers as low as possible thereafter, e.g. for example (3) the locants are 2,5- and not 3,6-.

Use of these names is subject to the following provisos:

(i) The rings are benzene (not hydrogenated).
(ii) There are not more than two methyl groups directly attached to the ring.
(iii) There is no alkyl substitution on the CH_3 groups.
(iv) There are no further phenyl groups attached directly to a benzene ring or to the nitrogen atom of aniline.

If any of these conditions is not met, go to section 21.9.2.

The following are retained as a basis for naming derivatives with the additional proviso that there is no alkyl substitution on the side-chain carbon atoms; if there is, aniline is the PFS†:

† Defined in section 1.5.

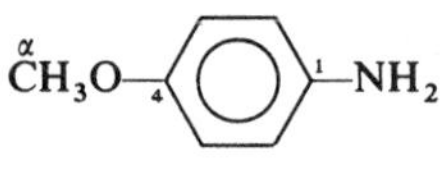

Anisidine
[*p*-isomer shown]

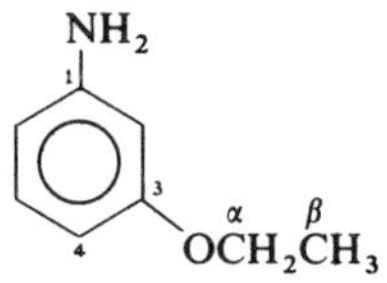

Phenetidine
[*m*-isomer shown]

21.9.2 Other monoamines

21.9.2.1 Primary

RNH_2

The name is formed by citing the radical R- (see Appendix C) and adding, without any space, the ending 'amine'.

Example (1)

$CH_3CH_2CH_2CH_2CH_2—NH_2$

Pentylamine

Example (2)

$$H_3C—C(CH_3)_2—CH_2—NH_2$$

Neopentylamine

Example (3)

$$CH_3—CH_2—CH(CH_2—CH_3)—NH_2$$

1-Ethylpropylamine

Example (4)

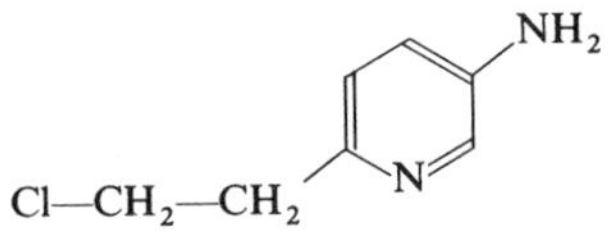

6-(2-Chloroethyl)-3-pyridylamine

Exceptions

If the radical is phenyl, whether or not substituted, see section 21.9.1. However, an intervening carbon-atom chain, however short, removes this provision, so

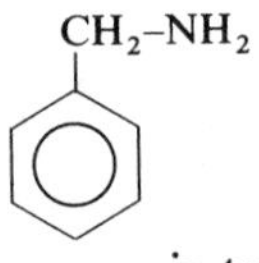

is treated here and is called benzylamine.

21.9.2.2 *Secondary*

(a) Symmetrical R$_2$NH

If the radical R is simple, i.e. unsubstituted, the name is formed by citing in order: 'di', then the name of the radical (see Appendix C) with no space, and then 'amine' (again with no space).

Example (1)

$$CH_3CH_2—NH—CH_2CH_3$$

Diethylamine

Example (2)

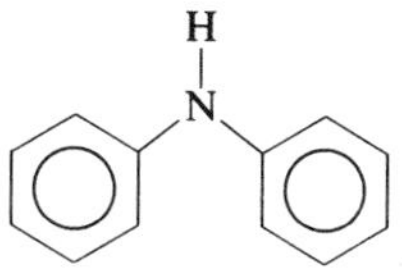

Diphenylamine

(NB. No embargo on 'phenyl' in this case)

If (usually because of being a contracted form) the radical-name begins with a numeral locant (to signify the position of attachment, for example), a hyphen is inserted in the name both before and after the numeral.

Example (3)

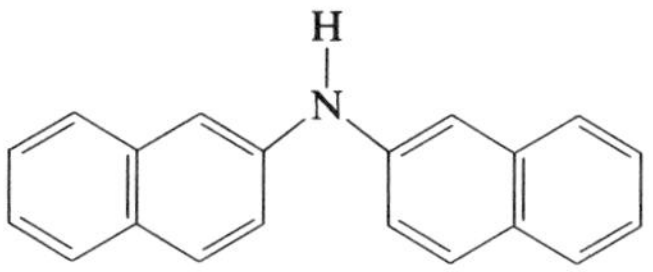

Di-2-naphthylamine

Example (4)

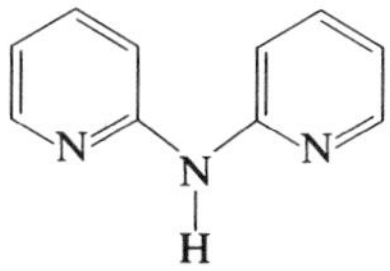

Di-2-pyridylamine

If the radical R is substituted, the name begins with 'bis' instead of 'di', and then follows (with neither space nor hyphen) the name of the compound radical enclosed in parentheses, and finally 'amine'.

Example (5)

$$CH_3—\overset{\underset{\displaystyle Cl}{|}}{CH}—CH_2CH_2—NH—CH_2CH_2—\overset{\underset{\displaystyle Cl}{|}}{CH}—CH_3$$

Bis(3-chlorobutyl)amine

Example (6)

Bis(2-iodopyrimidin-4-yl)amine

Example (7)

Bis(4-nitrophenyl)amine

(b) Non-symmetrical R^1R^2NH

Here R^1 and R^2 are different radicals. This can mean that they have the same skeletal structure differently substituted, or the same structure attached otherwise than at the same site. If one or both of the radicals is phenyl or a substituted phenyl-group, it may be excluded from consideration under this section and might be covered under mono-PG ring-systems (section 21.9.1). Therefore read that section and return here only if so directed.

The name ends in 'amine' but the names of the two dissimilar radicals are first separately cited (still with no spaces anywhere in the name) in alphabetical order. This introduces a complication because the name cyclopropylmethylamine could, in principle, cover two possible structures:

(8)   —CH₂—NH₂,   and   *(9)*   —NH—CH₃

In the case of structure (9) this ambiguity is overcome by enclosing the second-cited radical in parentheses; or, if it should have parentheses already, the next higher order of enclosing marks. Thus structure (9) becomes cyclopropyl(methyl)amine (retaining the alphabetical order of radicals), and structure (8) could be unambiguously conveyed as (cyclopropylmethyl)amine.†

This is not a hard-and-fast rule because sometimes no ambiguity exists in the straightforward citation of the radicals (as in example (4) of section 21.9.2.3 for instance).

Example (10)

—NH——OCH₂CH₃

Cyclobutyl(3-ethoxycyclobutyl)amine

† *N*-locants have also been used to overcome this difficulty and their use is well understood. However, they detract from the parent-hydride basis of the naming-scheme used here and are not recommended for this section.

Example (11)

1-Naphthyl-(2-naphthyl)amine
(note order of citation)

21.9.2.3 *Tertiary*

$$R^1R^2R^3N; \quad R^1R^2N; \quad R_3N$$

Where all three radicals are different (which includes the same structure differently substituted or attached), their names are cited in alphabetical order before 'amine'. Again, there are no spaces and hyphens are used only when they follow a numeral-locant or a Greek letter signifying a side-chain atom, as in benzyl. In this type of tertiary amine it may be necessary to use enclosing marks around each of the second and third cited radicals to avoid potential ambiguity.

Example (1)

Cyclopropyl(propyl)(pyrimidin-2-yl)amine
(order: c,pr,py)

Example (2)

2-Bromomethyl-3-methylcyclopropyl(2-bromopropyl)(pyrimidin-2-yl)amine
(Order: bromom,bromop,p)

If two or all three of the three radicals attached to the central nitrogen atom are identical and attached at the same position, they are collected in the name by di or tri, as the case may be—or bis or tris in the case of substituted radicals. Di and tri are followed by hyphens only if the radical begins with a numeral or N, α or β, whereas bis and tris are followed directly by open parentheses. If they occur unsubstituted or identically substituted twice or three times, phenyl groups may be included in this set, but the possible claims of section 21.9.1 should be examined and eliminated first.

Example (3)

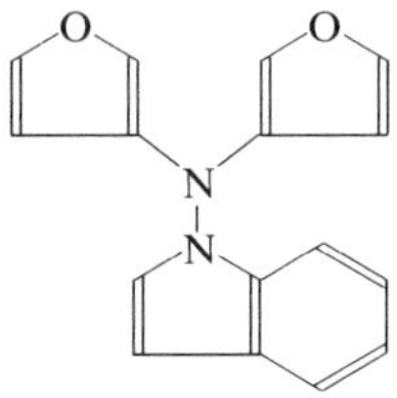

Di-3-furyl(indol-1-yl)amine

Example (4)

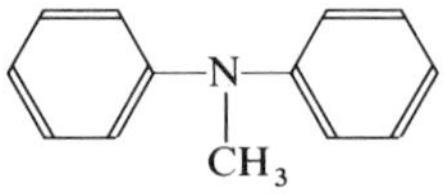

Methyldiphenylamine

It is argued that here no parentheses are needed because, if the methyl group were on one of the rings, that ring would be called 'tolyl'.

Example (5)

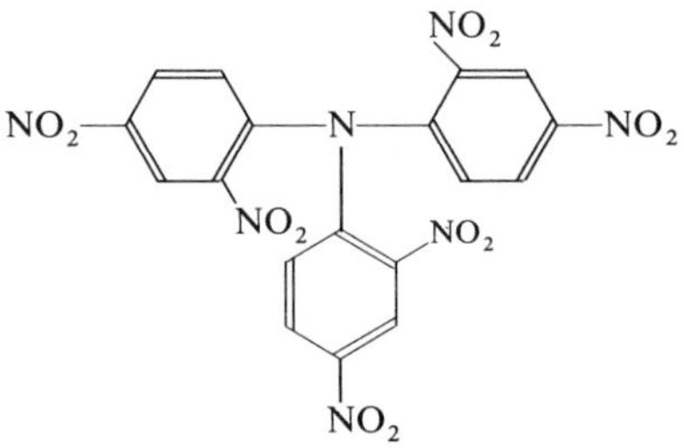

Tris(2,4-dinitrophenyl)amine

Example (6)

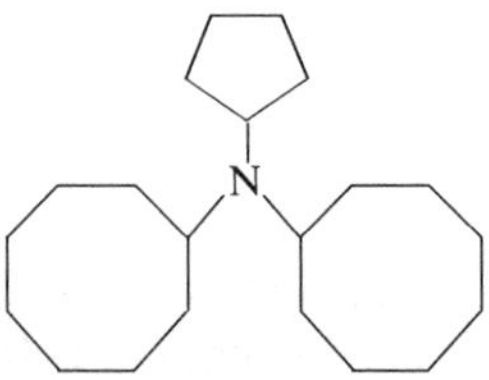

Di(cyclo-octyl)(cyclopentyl)amine

Example (7)

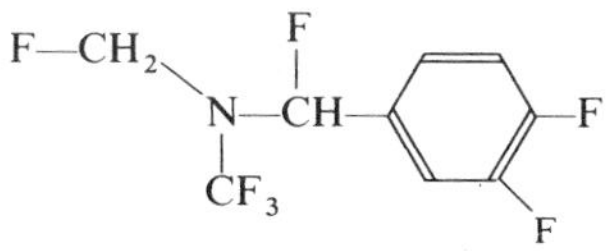

F—CH₂ \
N—CH— ...F \
CF₃ ... F F

Fluoromethyl(α,3,4-trifluorobenzyl)trifluoromethylamine

21.9.3   Diamines

21.9.3.1   Trivially named phenyl diamines

The following structure:

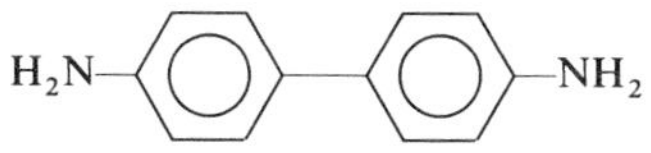

H₂N—⟨O⟩—⟨O⟩—NH₂

is named benzidine and its use as PFS is subject to the following provisos:

(a) it is used only for the isomer shown,
(b) there is no hydrogenation of either ring, and
(c) no other —N< groups are attached directly to either ring.

In the event of any hydrogenation, the criteria of the yellow pages, section A, decide the PFS (see Appendix D-1.1). For example, in case of hydrogenation of one ring only, the name would be based on the other, viz. aniline.

For other isomers, the name reverts to 'biphenyl-*x*, *y*-ylenediamine' where *x* and *y* are the locants for the —N< groups. This applies also to other polyamines of biphenyl, e.g. 'biphenyl-*x*, *y*,*z*-triyltriamine'.

The senior ring, as determined by yellow pages criteria (A-20)–(A-22) has its ring-locants and nitrogen atom 'unprimed', whilst those of the other ring are 'primed'.

Other cyclic polyamines are subdivided as shown in the following sub-sections.

A benzene ring having two amino-groups is named *o*, *m*, or *p*-phenylenediamine for the 1,2-, 1,3- and 1,4-isomers, respectively.

Example (1)

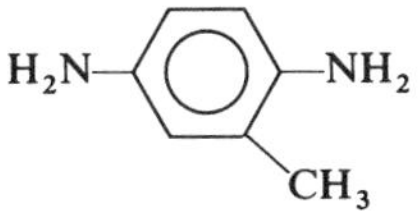

H₂N—⟨O⟩—NH₂ \
CH₃

2-Methyl-*p*-phenylenediamine
[NB. There is no 'toluylene' radical]

21.9.3.2   *Other collectable* diamines*

When both amino-groups are primary or else symmetrically N-substituted, the name is formed by citing the name of the appropriate divalent radical† separating the two N-atoms and adding "diamine".

Example (1)

$$NH_2CH_2CH_2CH_2NH_2$$

Trimethylenediamine

If the N-atoms are symmetrically substituted, the names for their attached groups (see Appendix C) are cited before the name of the parent diamine, in alphabetical order, each preceded by *N-* or *N'-*.
Composite radicals are enclosed in parentheses.

Example (2)

$$CH_3CH_2NH{-}CH_2CH_2CH_2NH{-}CH_2CH_3$$

N,N'-Diethyltrimethylenediamine

Example (3)

N,N,N',N',2,6-Hexamethyl(*p*-phenylenedimethylene)diamine

21.9.3.3   *Collectable diamines not symmetrically N-substituted*

The name is based on a senior mono amine, tertiary being preferred to secondary, and secondary to primary.

When both are tertiary, $NR^1R^1R^2$ is preferred to $NR^1R^2R^3$.

If a choice persists, follow the method of section 21.9.4.4, N-locants being reserved for the named structure of 21.9.1. For non-collectable (disconnected) diamines, e.g. $NH_2CH_2-O-CH_2CH_2NH_2$, see under section 21.9.4.4.

---

† This means that, whether or not it bears substituent groups, the central di- or poly-valent radical must be capable of being named in a way which depicts that it proceeds from a centre outwards simultaneously in two or more directions, e.g.

methylene $\leftarrow CH_2 \rightarrow$, nitrilo $\leftarrow N \rightarrow$, *o*-phenylenedioxy

but not in one direction only, e.g. oxymethylamino $\rightarrow O-CH_2 \rightarrow NH \rightarrow$.

*"Collected" defined in section 1.5

21.9.3.4 Guanidines

The compound

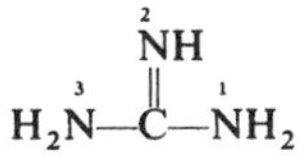

$$\overset{2}{N}H$$
$$\underset{}{\overset{\|}{C}}$$
$$H_2N\overset{3}{-}C\overset{1}{-}NH_2$$

is named 'guanidine' and numbered as shown.

In the absence of more senior FGs from Table 1, any attached groups are cited in alphabetical order as prefixes (see Appendix C): di, tri; or bis, tris, etc., are used to collect like groups, and the name ends with 'guanidine'. (1 and 3 are interchangeable and their choice depends on substitution, e.g. 1,1,3 is preferred to 1,3,3).

Example (1)

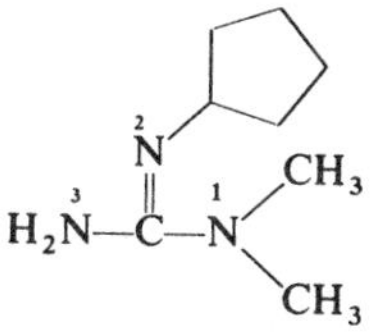

2-Cyclopentyl-1,1-dimethylguanidine

If any FGs from Table 1 are present, the senior PG determines the form of the name and the guanidine-unit is cited as a prefix depending on the position by which it is attached. These prefixes are respectively:

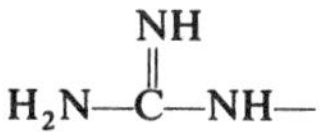

Guanidino   and   Diaminomethyleneamino

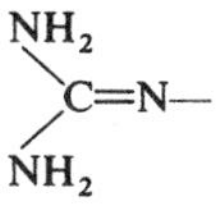

21.9.4 Polyamines with three or more nitrogen atoms

21.9.4.1 All primary with two or more collectable

As many amines as possible are combined in the name-ending by means of the appropriate polyvalent radicals. (see Appendix C). Those for which this is impossible are cited as amino-prefixes.

Example (1)

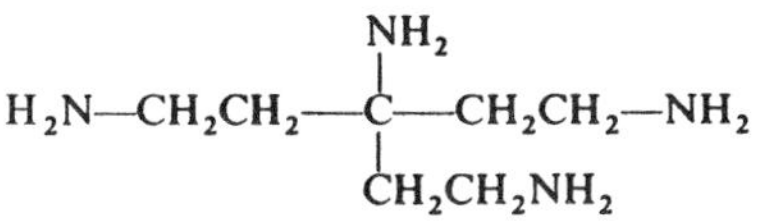

3-(2-Aminoethyl)pentane-1,3,5-triyltriamine

21.9.4.2 One or more secondary or tertiary

The name which puts as many primary amines as possible into the name-ending is chosen—provided that there are more than one. Where this is not the case, the procedures of section 21.9.4.4 are followed.

Example (1)

$$H_2N—CH_2—CH—CH_2—NH_2$$
$$CH_3—N—CH_3$$

2-Dimethylaminotrimethylenediamine

21.9.4.3 Individual chains with four or more non-terminal, non-contiguous nitrogen atoms

The chain is numbered from one terminal carbon atom to the other and the appropriate polyvalent radical attached to primary amine groups (so long as these do not exceed two) cited, before the ending triamine, tetramine, etc., as the case may be. The N-positions in the chain are cited immediately before the name of the chain as diaza, triaza, etc., prefixes according to section 4 (q.v.). The name begins with radical prefixes for attached substituents, cited in alphabetical order. If a single chain carries more than two primary amine groups, the name is formed according to section 21.9.4.1.

Example (1)

$$H_2N—\overset{1}{C}H_2—\overset{2}{N}\overset{CH_3}{\diagup} \qquad \overset{CH_3}{\diagup}$$
$$\overset{3}{CH_2}—\overset{4}{NH}—\overset{5}{CH_2}—\overset{6}{NH}—\overset{7}{CH_2}—\overset{8}{N}\diagup \quad \underset{CH_2CH_2NH_2}{\diagdown}$$

2,8-Dimethyl-2,4,6,8-tetra-azadecamethylenediamine

If there is only one primary amine, the main chain attached to it is cited as a radical prefix (still following section 4 procedures) and the name ends in 'amine'.

Example (2)

$$NH_2—\overset{1}{C}H_2—\overset{CH_3}{\underset{2}{N}}—\overset{3}{C}H_2—\overset{4}{NH}—\overset{5}{C}H_2—\overset{CH_3}{\underset{6}{N}}—\overset{7}{C}H_2\overset{8}{C}H_2—\overset{9}{N}\overset{CH_3}{\diagup}\overset{10}{\diagdown}CH_3$$

2,6,9-Trimethyl-2,4,6,9-tetra-azadecylamine

21.9.4.4 Non-collectable* polyamines (including diamines)

The parent amine is chosen by considering all the amino-groups present and giving priority to any of the set of trivially named phenylamines from 21.9.1 that are present in the priority-

*"Collected" defined in section 1.5

order: phenetidine; anisidine; xylidine; toluidine; aniline (subject to their accompanying provisos, q.v.).

If none of these is appropriate, then tertiary is preferred to secondary, and secondary to primary. Then R_3N is preferred to $R^1{}_2R^2N$ and R^2N is preferred to R^1R^2N. If a choice remains, criterion B(15) Yellow pages is applied.

If section 21.9.1 is not appropriate, the name is formed as in section 21.9.2, the non-preferred groups being cited in alphabetical order as radical prefixes (see Appendix C).

Example (1)

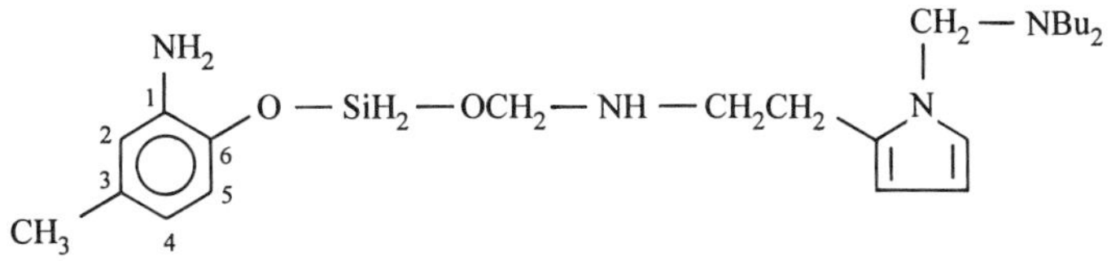

6-{2-[1-(Dibutylaminomethyl)pyrrol-2-yl]ethylaminomethoxysiloxy}-*m*-toluidine

Example (2)

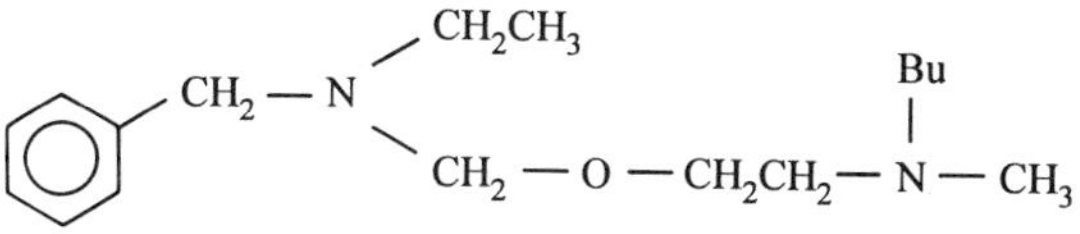

(i)   Benzyl{2-[butyl(methyl)amino]ethoxymethyl}ethylamine
(ii)   2-[Benzyl(ethyl)aminomethoxy]ethyl(butyl)(methyl)amine
[(i) preferred to (ii) by B(15) Yellow pp.: benzyl b rather than benzyl e.
Note the use of enclosing marks to avoid ambiguity.]

21.10   IMINES

$$R{=}NH; \qquad R^1{=}N{-}R^2$$

Imines of the first type are named by adding 'amine' to the name of the appropriate divalent radical.

Example (1)

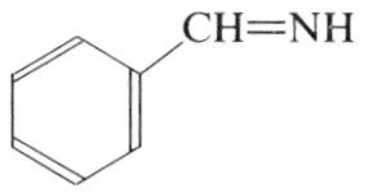

Benzylideneamine
[For 'benzylidene', see Appendix C-1.2.]

Example (2)

$$Cl{-}CH{=}CH{=}NH$$

2-Chlorovinylideneamine

Imines of the second type are named in a manner analogous to that for diamines formed by substitution into 'ammonia' of two dissimilar radicals. Thus, the name of each radical is cited in turn, in alphabetical order, regardless of valency, the monovalent radical ending in -yl and the divalent radical ending in -ylidene. The name ends with 'amine' and there are no spaces. Hyphens follow numerals but are not otherwise used. If phenyl is attached directly to the nitrogen atom, consult section 21.9.1. Parentheses are used where necessary to avoid ambiguity.

Example (3)

$$CH_3 \text{—}\triangle\text{—} N{=}CH_2$$

2-Methylcyclopropyl(methylene)amine

Example (4)

$$CH_3CH_2\text{—}N{=}CH\text{—}CH_3$$

Ethyl(ethylidene)amine

Example (5)

Cyclohexylidene(2*H*-pyran-4-yl)amine

21.11 ETHERS

$$R\text{—}O\text{—}R; \quad R^1\text{—}O\text{—}R^2$$

21.11.0 General

The term 'ether' is applied to compounds characterized by a PG or solitary group consisting of an oxygen atom linked directly to two other groups, attached at a carbon atom (and no other) in each case. The two groups may be identical or not. ['Identical' here means the same groups attached in the same manner; thus 1-naphthyl is not identical with 2-naphthyl].

21.11.1 Trivially named ethers

The following compounds have trivial names which can serve as a basis for naming substituted derivatives (provided no group senior to ether is present), and these are used in preference to the radicofunctional alternatives in the following sub-sections.

Example (1)

$$\bigcirc\text{—}O\text{—}\overset{\alpha}{C}H_3$$

Anisole

Example (2)

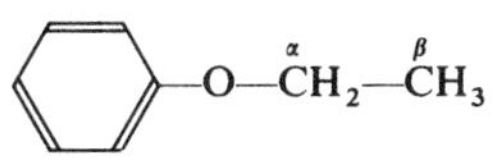

Phenetole

Example (3)

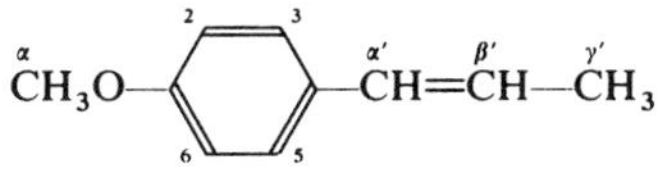

Anethole

Note that here the ring is numbered with arabic numerals whilst Greek letters are used for the side-chains.

If any of the side-chain atoms are linked directly to one or more unfused benzene rings or open-chain carbon atoms, or if an unfused benzene ring is joined directly to the ring, the trivial name is dropped in favour of the three-word style, ending in 'ether' (see examples (4–6) of section 21.11.2).

For anethole to be retained in the name the double bond must be present in the three-carbon chain. α',β'-Dihydro derivatives are also named by the three-word method.

Example (4)

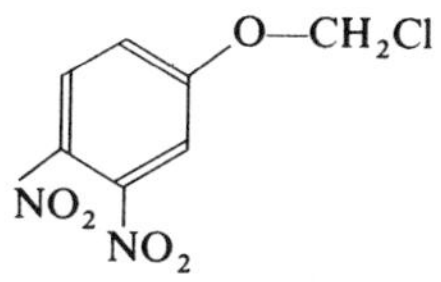

α-Chloro-3,4-dinitroanisole

Example (5)

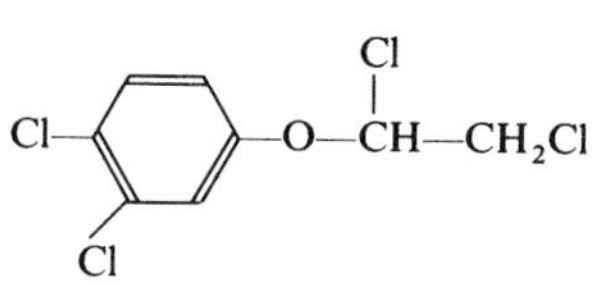

$\alpha,\beta,3,4$-Tetrachlorophenetole

Example (6)

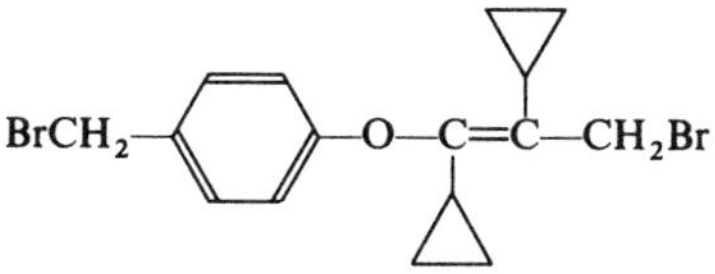

α,γ'-Dibromo-α',β'-dicyclopropylanethole

Example (7)

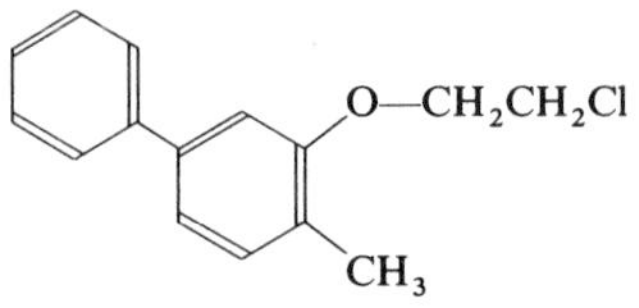

2-Chloroethyl 4-methylbiphenyl-3-yl ether
(not β-chloro-2-methyl-5-phenylphenetole)

Example (8)

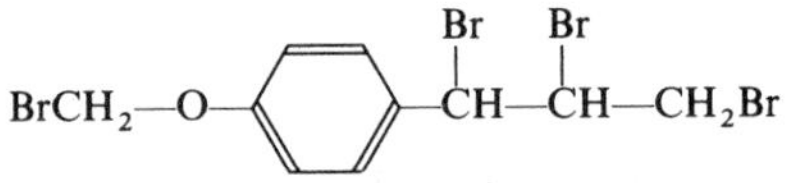

Bromomethyl 4-(1,2,3-tribromopropyl)phenyl ether
(not α,α′,β′,γ′-tetrabromo-α′,β′-dihydroanethole)

21.11.2 Mono-ethers, systematic radicofunctional names

If the two attached groups are the same, the name begins with di or bis. As above, bis is used either (a) when the group carries one or more substituents, or (b) to avoid ambiguity consequent on the use of di. The radical name follows, then a space, then the word 'ether'.

Example (1)

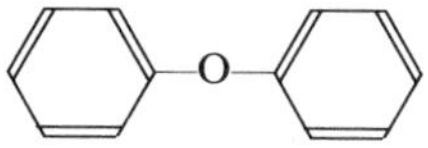

Diphenyl ether

Example (2)

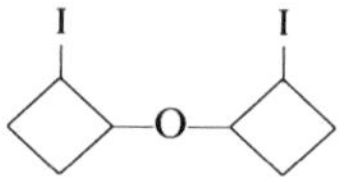

Bis(2-iodocyclobutyl) ether

Example (3)

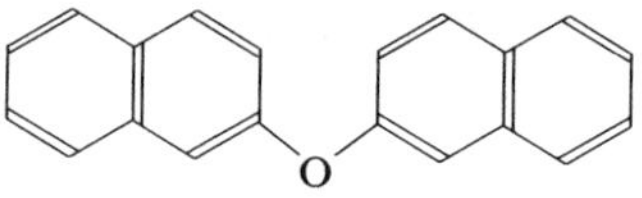

Di-2-naphthyl ether

If the two groups are dissimilar (and that includes dissimilar substitution of the same structure and attachment by a different site of the same structure), they are cited in alphabetical order and separated by a space. After another space, comes the word 'ether', so this type are named using three words.

Example (4)

$$CH_3CH_2—O—CH=CH_2$$

Ethyl vinyl ether

Example (5)

1-Anthryl 2-bromoethyl ether

Example (6)

2-Pyridyl pyrimidin-2-yl ether

If the two groups (whether identical or not) are attached to each other by any linkage in addition to the etheric –O–, they are named under the sections for heterocyclic or bridged systems (9 or 19.1).

21.11.3   Di- and poly-ethers

 (i) For di-ethers, each oxygen atom is considered in turn as the class-name site and, of the two possible names, that coming earliest in the alphabet is preferred.
 (ii) For tri-ethers the central oxygen atom is used as the basis of the name and its flanking-group names cited in alphabetical order as for a mono-ether.
(iii) For higher numbers of oxygen atoms, section 4 should be considered but, for example, in symmetrical cases, the methods of (i) or (ii) are followed (using 'bis') as appropriate.

Example (1)

$$CH_3CH_2—O—CH_2CH_2—O—CH_2CH=CH_2$$

Allyl 2-ethoxyethyl ether
[not 2-allyloxyethyl ethyl ether: allyl-e- preferred to allyl-o-]

Example (2)

$$CH_3CH_2—O—CH_2—O—CH_2—O—CH_2CH_3$$

Bis(ethoxymethyl) ether

Example (3)

$$C_6H_5—O—CH_2CH_2—O—CH_2—O$$
$$C_6H_5—O—CH_2CH_2—O—CH_2—O \diagdown CH_2$$

1,11-Diphenoxy-3,5,7,9-tetraoxaundecane

21.12 SULPHIDES AND POLYSULPHIDES

$$(R^1—[S]_x—R^2)$$

21.12.1 Monosulphides

These are named in a manner strictly analogous to ethers except that there are no trivial names requiring exceptional treatment.

Instead of the word 'ether' their names end in 'sulphide'. Apart from that, the same considerations hold good as for ethers (section 21.11, q.v).

Example (1)

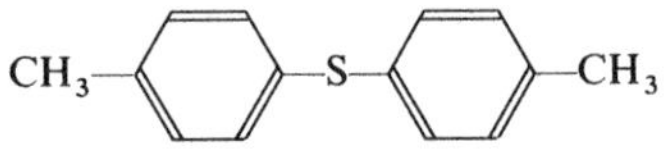

Di-*p*-tolyl sulphide

Example (2)

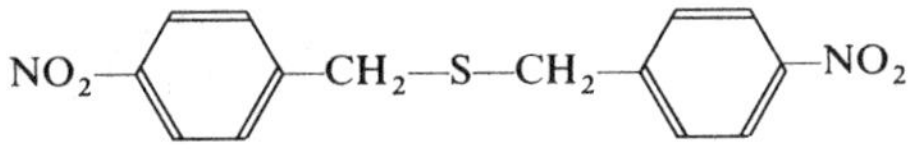

Bis(4-nitrobenzyl) sulphide

Example (3)

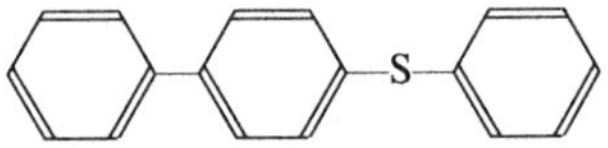

Biphenyl-4-yl phenyl sulphide

21.12.2 Polysulphides

$$(R'—(S)_x—R)$$

R and R' may be the same or different, and $x > 1$.

When $x = 2$, the name ends in 'disulphide'; when $x = 3$, it ends in 'trisulphide', etc. Otherwise, name-construction is exactly as for sulphides.

Example (1)

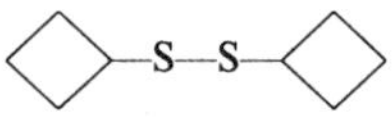

Dicyclobutyl disulphide

Example (2)

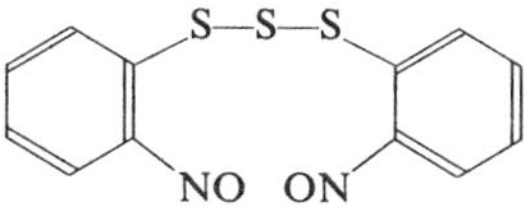

S—S—S

Bis(2-nitrosophenyl) trisulphide

Example (3)

$$Cl—CH=CH—S—S—S—S—CH=CH_2$$

2-Chlorovinyl vinyl tetrasulphide

21.13 SULPHOXIDES

$$\overset{\displaystyle O}{\underset{\displaystyle (R—S—R)}{\uparrow}}$$

R stands for a group attached by a single C–S bond. The two R-groups may be the same or different.

Names are constructed as for sulphides except that they end not in 'sulphide' but in 'sulphoxide'.

Example (1)

$$\overset{\displaystyle O}{\underset{\displaystyle CH_3—S—CH_3}{\uparrow}}$$

Dimethyl sulphoxide

sometimes depicted as

$$\overset{\displaystyle O}{\underset{\displaystyle CH_3—S—CH_3}{\|}}$$

Example (2)

$$CH_3CH_2—\overset{\displaystyle O}{\underset{}{\overset{\|}{S}}}$$

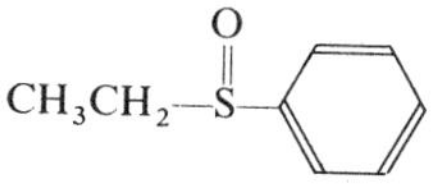

Ethyl phenyl sulphoxide

21.14 SULPHONES

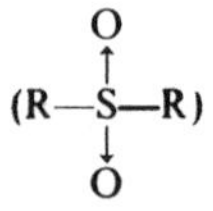

$$(R-\overset{\uparrow}{\underset{\downarrow}{\overset{O}{\underset{O}{S}}}}-R)$$

(sometimes depicted as $\quad R-\overset{O}{\underset{O}{\overset{\|}{S}}}-R\quad$ or $\quad R-SO_2R$)

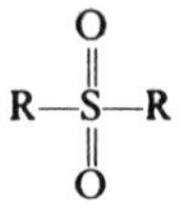

Exactly the same considerations apply here as for sulphoxides, except that the name ends in 'sulphone'.

Example (1)

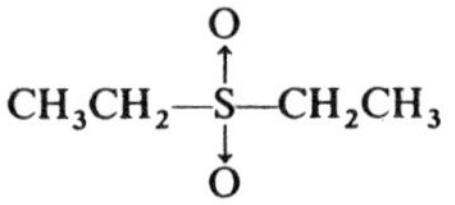

$$CH_3CH_2-\overset{\uparrow}{\underset{\downarrow}{\overset{O}{\underset{O}{S}}}}-CH_2CH_3$$

Diethyl sulphone

Example (2)

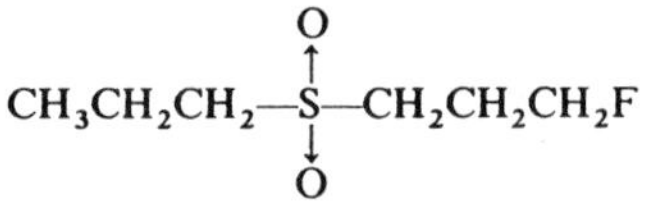

$$CH_3CH_2CH_2-\overset{\uparrow}{\underset{\downarrow}{\overset{O}{\underset{O}{S}}}}-CH_2CH_2CH_2F$$

3-Fluoropropyl propyl sulphone

21.15 PEROXIDES

$$(R-O-O-R;\ R^1-O-O-R^2)$$

These are named in a manner analogous to that for ethers except that there are no trivial names requiring exceptional treatment. The names end in 'peroxide' but, apart from that, name-construction is as for ethers (see section 21.11).

Example (1)

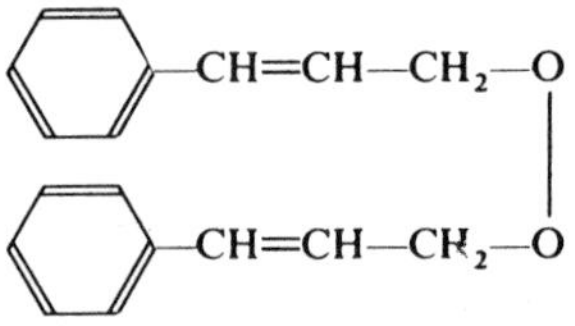

Dicinnamyl peroxide

Example (2)

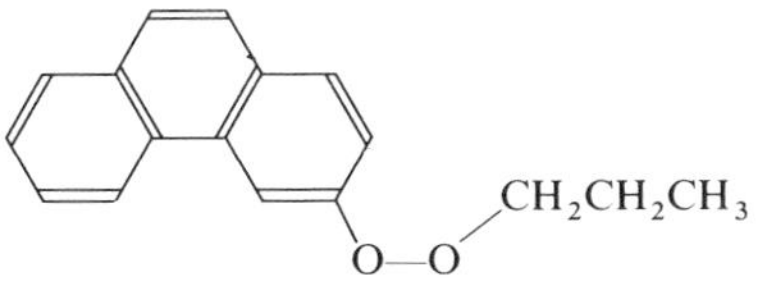

Phenanthren-3-yl propyl peroxide

This style of name is also used for acyl groups $(R\!-\!\overset{\displaystyle O}{\underset{\displaystyle \|}{C}}\!-)$ on both sides of the —O—O—group. (Note that when there is an acyl-group on one side only, it is not a peroxide but an ester.)

Example (3)

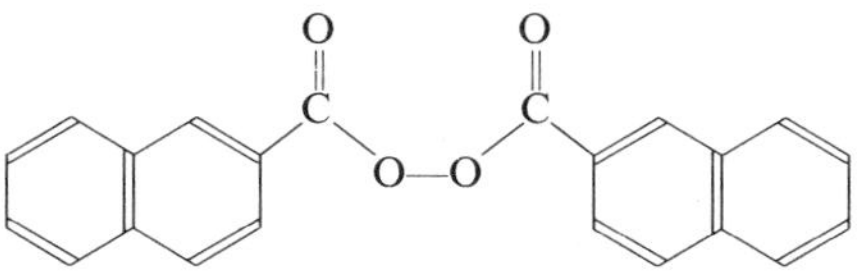

Di-2-naphthoyl peroxide

Example (4)

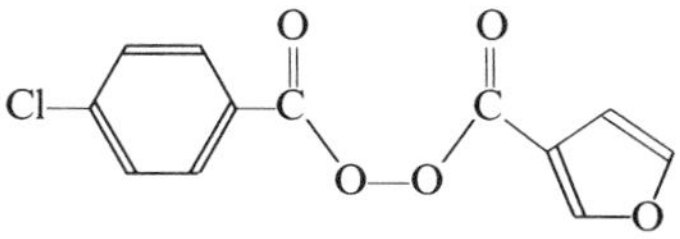

4-Chlorobenzoyl 3-furoyl peroxide

21.16 AZIDES

$$(R\!-\!N_3;\quad R\!-\!\overset{\displaystyle O}{\underset{\displaystyle \|}{C}}\!-N_3)$$

As indicated, the radical may be any 'carbon-attached' radical, including acyl radicals. The name is formed by citing the name of the radical, then, after a space, the word 'azide'.

Example (1)

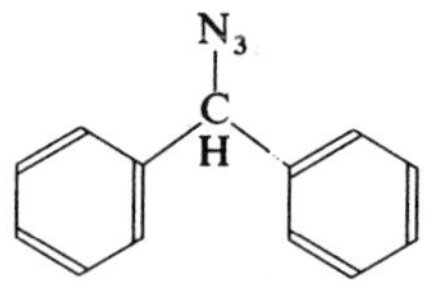

Benzhydryl azide

Example (2)

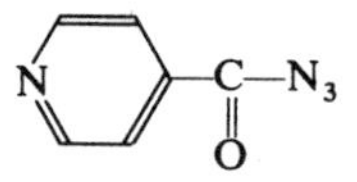

Isonicotinoyl azide

21.17 OXIDES

When a heteroatom, such as sulphur or nitrogen, carries one or two $=O$ groups whilst occupying a ring-position, the compound is named by first removing such $=O$ groups and then addressing the resultant structure to the **flow-diagram** at the back of the book. After the name (derived according to the appropriate sectional instructions) there follows a space and then 'oxide' or 'dioxide' as the case may be, preceded by appropriate locant-numerals or letters to indicate the element concerned.

Example (1)

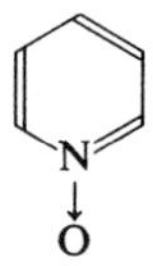

Pyridine 1-oxide

Example (2)

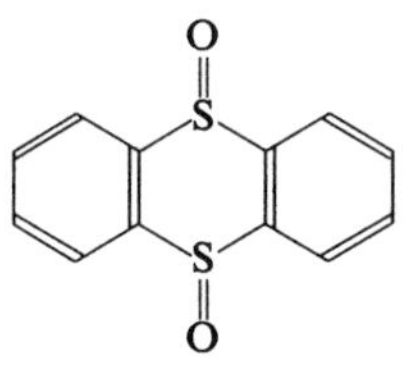

Thianthrene 5,10-dioxide

Example (3)

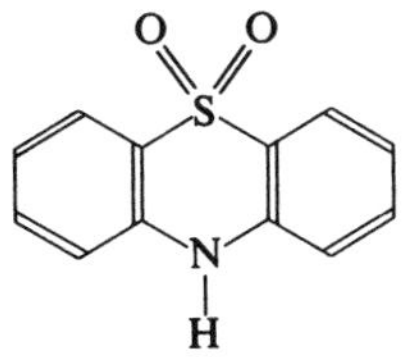

Phenothiazine 5,5-dioxide

Example (4)

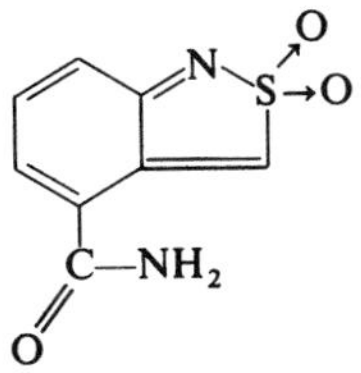

2,1-Benzothiazole-4-carboxamide 2,2-dioxide

22 Vari-functional acyclic compounds†

This section lists the trivial names which cover not only a characteristic chain and its attached PG (see definitions section 1.5) but also subsidiary functional groups (FG). Their use for naming substituted derivatives is preferred over more systematic names and they are regarded as senior, in the event of a choice of parent name, to chains carrying the same PG in the same numbers but without the subsidiary FG.

To reach the appropriate section, apply your structure to the subsidiary flow-diagram of section 22.0.

---

† 'The term 'acyclic' is used here in the sense of PGs not being directly attached to a ring or ring-system.

22.0   SUBSIDIARY FLOW-DIAGRAM

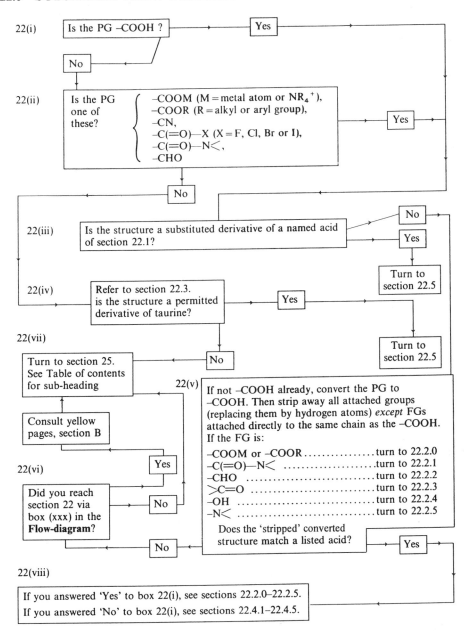

22.1   CARBOXYLIC ACIDS WITH TRIVIAL NAMES COVERING MORE THAN ONE KIND OF SUBSIDIARY FUNCTIONAL GROUP

The following trivial names are retained for use as PFS‡ in naming substituted derivatives (see section 22.5).

‡ PFS—see Appendix- D-1.1 & section 1.5.

For examples (1–4), (11) and (12) of this subsection the name is retained for O-acyl substituents but, in the case of other O-substitution, the –OR or –OAr groups are named according to Appendix C and the parent structure is named as though this –O– group was absent.

Example (1)

$$\overset{3}{HO-CH_2}-\overset{2}{CH(NH_2)}-\overset{1}{COOH}$$

Serine

Example (2)

$$\overset{4}{CH_3}-\overset{3}{CH(OH)}-\overset{2}{CH(NH_2)}-\overset{1}{COOH}$$

Threonine

Example (3)

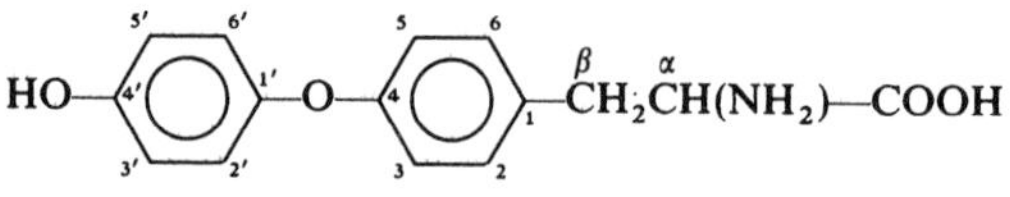

Thyronine

Example (4)

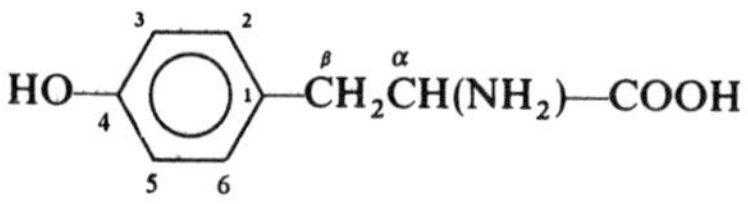

Tyrosine

Example (5)

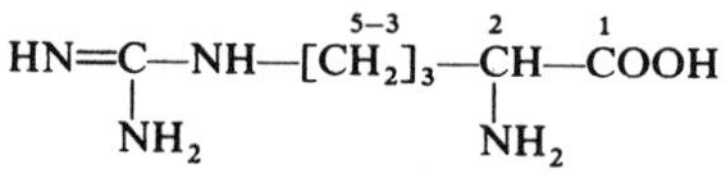

$$HN=\overset{}{C}-NH-[\overset{5-3}{CH_2}]_3-\overset{2}{CH}-\overset{1}{COOH}$$
$$\underset{NH_2}{\quad\quad} \quad\quad\quad\quad \underset{NH_2}{\quad}$$

Arginine

Example (6)

$$\overset{3}{HS-CH_2}-\overset{2}{CH(NH_2)}-\overset{1}{COOH}$$

Cysteine

Example (7)

$$\overset{4}{HS-CH_2}\overset{3}{CH_2}\overset{2}{CH(NH_2)}-\overset{1}{COOH}$$

Homocysteine§

---

§ The prefix 'homo' may be used with other amino acid trivial names to mean that a CH_2-group has been added to lengthen the carbon-chain, as here.

Example (8)

$$\overset{4}{CH_3}-S-\overset{3}{CH_2}\overset{2}{CH_2}\overset{}{CH}(NH_2)-\overset{1}{COOH}$$

Methionine

Example (9)

$$H_2N-\overset{4}{C}(=O)-\overset{3}{CH_2}-\overset{2}{CH}(NH_2)-\overset{1}{COOH}$$

Asparagine

Example (10)

$$H_2N-\overset{5}{C}(=O)-\overset{4}{CH_2}\overset{3}{CH_2}\overset{2}{CH}(NH_2)-\overset{1}{COOH}$$

Glutamine

Example (11)

$$H_2N-\overset{4}{C}(=O)-\overset{3}{CH}(OH)-\overset{2}{CH}(OH)-\overset{1}{COOH}$$

Tartaramic acid

Example (12)

$$\overset{4}{HC}(=O)-\overset{3}{CH}(OH)-\overset{2}{CH}(OH)-\overset{1}{COOH}$$

Tartaraldehydic acid

Oligopeptides

Although peptides are excluded from consideration (see Appendix A), cases involving only a few amino acids of the above set, or those of sections 22.2.5 or 26.1.1, may also occur, in which each −NH unit of one acid is joined directly to the acyl [−C(=O)] group of the next in a sequential arrangement to give:

$$H_2N- \dots -CONH- .. -[CONH]_x \dots COOH.$$

These are named by citing the radical prefix form appropriate to the amino acid at the extreme left-hand end of this presentation, then that of its neighbouring acid, and so on, ending with the name of the right-hand end acid, e.g. lysylserylvaline. L-forms are assumed for naturally occurring amino acids if not inserted. If D-forms are appropriate, or if in doubt, insert them, flanked by hyphens, e.g. D-alanyl-D-aspartyl-L-serylglycine. For more complicated cases, consult CNAS.

22.2 TRIVIAL NAMES COVERING –COOH AND A FUNCTIONAL GROUP

22.2.0   –COOH plus either –COOM or –COOR: acid salts and acid esters

M is a metal atom or quaternary N^+ ion; R is an alkyl or aryl group.

The structure may be an acid salt or an acid ester. In the case of acid salts, 'replace' the M by a hydrogen and see if the two resultant –COOH groups can be 'collected' (see section 1.5) in the name. For this they must be attached to the same uninterrupted carbon-chain thus:

$$HOOC-[C]_x-COOH.$$

If this is the case, the name takes the form:

metal (or cation)—space—hydrogen—space—(....)anion.

If the 'stripped' structure matches a trivially named acid of section 26.1.1 or 26.1.2 (q.v.), the anion name is derived from that of the acid cited there; if not, a systematically named acid formed as under section 26.1.3 is used instead. In both cases the anion name is formed as described in section 26.1.5.

NB. If tartaric acid is applicable, the anion name is 'tartrate' and not 'tartarate'.

The (....) stand for appropriately named radical prefixes for any attached groups (other than those covered by trivial names of sections 22.2.1–22.2.5), cited without a break, in alphabetical order.

Example (1)

$$Na^+[HOOC-CF_2-COO]^-$$

Sodium hydrogen difluoromalonate

For salts of polyvalent metals, the 'hydrogen' part of the name is preceded by 'di' or 'tri' as appropriate, and the anion name is preceded by 'di' or 'tri' or, if substituted, by 'bis' or 'tris' and enclosed in parentheses.

Example (2)

$$Mg^{2+}[HOOC-CH(CH_3)-CHCl-COO^-]_2$$

Magnesium dihydrogen bis(2-chloro-3-methylsuccinate)

Example (3)

$$[CH_3]_4N^+[HOOC-[CH_2]_8-COO]^-$$

Tetramethylammonium hydrogen decanedioate

If the two –COO groups cannot be 'collected', the chain bearing the –COOH group is named according to section 25, (q.v.) and the –COOM group takes its alphabetical place in radical form among the substitutive prefixes (e.g. 'sodio-oxycarbonyl').

In the case of acid esters, two possible situations may apply:

(A)   ⌇COOH   (B)   ⌇COOH
       ⌇COOR   ⌇O—C(=O)—R

In case (A) the same considerations apply as for salts: if the –COO groups are 'collectable', the name takes the form:

radical–space–hydrogen–space–anion — the radical being R.

As for salts, the anion will be e.g. 'malonate', 'succinate', if it matches one of the trivially named set from sections 26.1.1 or 26.1.2 after the stripping process of box 22(v) in section 22.0 and the R→H conversion. If not, the acid will be systematically named in section 26.1.3 and the anion formed as in section 26.1.5.

If the −COO groups are not collectable, the name will be based on that −COOH group chosen, after consulting the yellow pages, as described in section 25.1.1 or 25.1.2, depending upon the case.

In case (B) the same situation applies, i.e. if the replacement of the —(C=O)—R by —H yields one of the trivially named hydroxy-acids of section 22.2.4, that provides the appropriate name-ending. If not, consult the yellow pages and then proceed to section 25.1.1 or 25.1.2 as appropriate.

22.2.1   —COOH plus —C(=O)—N< : amic acids

'Convert' the —C(=O)—N< group, complete with any N-substituents, to −COOH. Strip any other attached groups, except the FG, as per box 22(v) of section 22.0 and see if the two −COO groups can be 'collected' (see section 1.5). If not, turn to section 25.1, and see the yellow pages. If they are collectable (i.e. on the same uninterrupted carbon-chain), compare the stripped dicarboxylic acid with those of sections 26.1.1 and 26.1.2. If they match, the appropriate trivial name is converted from 'ic acid' to 'amic acid'.

Exception

NH$_2$—C(=O)—COOH is not called 'oxalamic acid' but 'oxamic acid'

If there is no such matching, turn to section 25.1.

In the case of *N*-phenyl derivatives, the 'ic acid' is replaced by 'anilic acid'.

Example (1)

$$^1COOH$$
$$|$$
$$^2CH_2$$
$$|$$
$$^3CONH_2$$

Malonamic acid

Example (2)

$$^2CH_2—COOH \quad ^1$$
$$|$$
$$_3CH_2—CONH_2$$
$$_4$$

Succinamic acid

Example (3)

$$^3CH_2—CH_2—COOH \quad ^2 \quad ^1$$
$$|$$
$$_4CH_2—CH_2—CONH—^{1'}\bigcirc_{4'}$$
$$_5 \quad _6$$

Adipanilic acid
[cf. section 25.14]

22.2.2 —COOH plus —CHO : aldehydic acids

'Convert' the –CHO to –COOH. If the resultant –COOH pair are collectable, compare the structure after box 22(v) of section 22.0 with the trivially named acids of sections 26.1.1 and 26.1.2. If there is a match, the appropriate acid name is converted from 'ic acid' to 'aldehydic acid'.

Example (1)

$$^3CHO$$
$$|$$
$$_2CH_2\!-\!\underset{1}{C}OOH$$

Malonaldehydic acid

Example (2)

$$\underset{}{HC}\!-\!\underset{1}{COOH}$$
$$\|$$
$$\underset{3}{HC}\!-\!\underset{4}{CHO}$$

Malealdehydic acid

Example (3)

$$HO\underset{1}{O}C\!-\![CH_2]_3\!-\!\underset{5}{C}HO$$

Glutaraldehydic acid

If there is no match with the trivially named acids, or if the –COO groups are not collectable, turn to section 25.1.

22.2.3 —COOH plus >C=O : keto acids (oxo acids)

The following trivially named acids are retained as PFSs (see Appendix D-1.1) provided their carbon-chains are not lengthened by alkyl substitution. If they are, the name is formed according to section 25.1, the =O groups being expressed in the name as 'oxo' prefixes. Numbering is as shown.

Example (1)

$$\underset{2}{HC}(=\!O)\!-\!\underset{1}{C}OOH$$

Glyoxylic acid

Example (2)

$$^3CH_3\!-\!\underset{2}{C}(=\!O)\!-\!\underset{1}{C}OOH$$

Pyruvic acid

Example (3)

$$^3CH_3\!-\!\underset{2}{C}(=\!O)\!-\!\underset{1}{C}H_2COOH$$

Acetoacetic acid

22.2.4 −COOH plus −OH : hydroxy acids

The following trivially named acids are retained as PFSs, provided their carbon-chains are not lengthened by alkyl substitution. In that event, the name is formed according to section 25.1, the −OH groups being expressed in the name as 'hydroxy' prefixes.

Example (1)

$$\overset{2}{\text{HO}}-\overset{1}{\text{CH}_2\text{COOH}}$$

Glycolic acid

Example (2)

$$\overset{3}{\text{CH}_3}\overset{2}{\text{CH}}(\text{OH})-\overset{1}{\text{COOH}}$$

Lactic acid

Example (3)

$$\text{HO}-\overset{3}{\text{CH}_2}-\overset{2}{\text{CH}}(\text{OH})-\overset{1}{\text{COOH}}$$

Glyceric acid

These names are still used when the −OH hydrogen atom is replaced by an acyl group (see Appendix C), but not when it is replaced by any other group, such as an alkyl or aryl group.

In such cases, the methods of section 25.1 decide the name, the compound oxy-radicals being formed as described in Appendix C (q.v.).

22.2.5 −COOH plus —N< : amino acids

The following trivial names are retained as PFSs (see section 1.5 & Appendix D-1.1). Aside from the permitted use of 'homo' (see footnote in section 22.1) for unambiguous cases, they may not be used where the carbon-chain has been lengthened by alkyl substitution.

In the event of a choice, consult the yellow pages to determine which of these acids to use for the name-ending.

Example (1)

$$\text{H}_2\text{N}-\text{COOH}$$

Carbamic acid

Example (2)

$$\overset{3}{\text{CH}_3}-\overset{2}{\text{CH}}(\text{NH}_2)-\overset{1}{\text{COOH}}$$

Alanine

Example (3)

$$\text{H}_2\text{N}-\overset{2}{\text{CH}_2}-\overset{1}{\text{COOH}}$$

Glycine

Example (4)

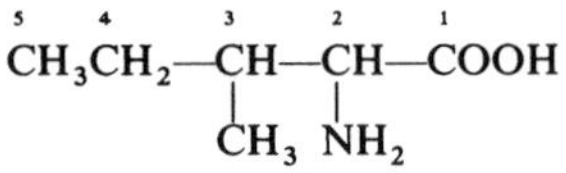

$$\overset{5}{C}H_3\overset{4}{C}H_2\!-\!\overset{3}{C}H\!-\!\overset{2}{C}H\!-\!\overset{1}{C}OOH$$
$$\qquad\qquad\;\;|\qquad\;|$$
$$\qquad\qquad\;CH_3\;\;NH_2$$

Isoleucine

Example (5)

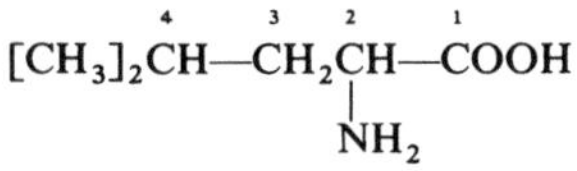

$$[CH_3]_2\overset{4}{C}H\!-\!\overset{3}{C}H_2\overset{2}{C}H\!-\!\overset{1}{C}OOH$$
$$\qquad\qquad\qquad\quad|$$
$$\qquad\qquad\qquad\;NH_2$$

Leucine

Example (6)

$$H_2N\!-\![\overset{6-3}{C}H_2]_4\!-\!\overset{2}{C}H\!-\!\overset{1}{C}OOH$$
$$\qquad\qquad\qquad\quad|$$
$$\qquad\qquad\qquad\;NH_2$$

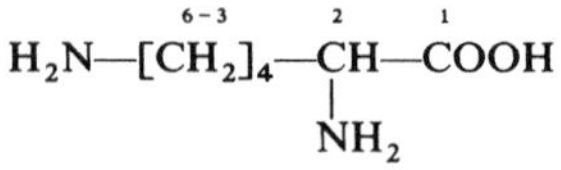

Lysine

Example (7)

$$\overset{6}{C}H_3\!-\![\overset{5-3}{C}H_2]_3\!-\!\overset{2}{C}H(NH_2)\!-\!\overset{1}{C}OOH$$

Norleucine

Example (8)

$$\overset{5}{C}H_3\overset{4}{C}H_2\overset{3}{C}H_2\overset{2}{C}H(NH_2)\!-\!\overset{1}{C}OOH$$

Norvaline

Example (9)

$$H_2N\!-\![\overset{5-3}{C}H_2]_3\overset{2}{C}H(NH_2)\!-\!\overset{1}{C}OOH$$

Ornithine

Example (10)

$$CH_3\!-\!NH\!-\!\overset{2}{C}H_2\!-\!\overset{1}{C}OOH$$

Sarcosine *

Example (11)

$$\overset{4}{(C}H_3)_2\overset{3}{C}H\!-\!\overset{2}{C}H(NH_2)\!-\!\overset{1}{C}OOH$$

Valine

Example (12)

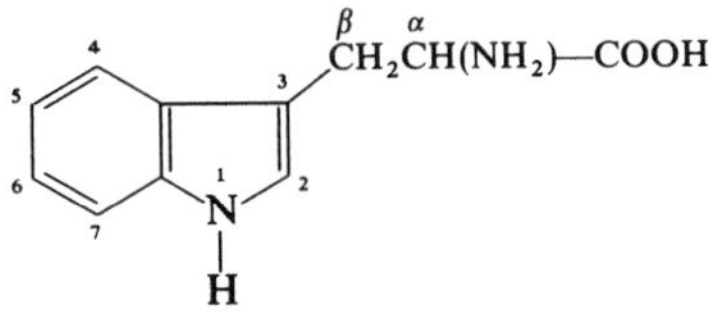

Tryptophan

*If the CH$_3$-group has any substituents the name is based on glycine.

Example (13)

$$CH_2CH(NH_2)-COOH$$

Histidine

Example (14)

$$C-NH-CH_2-COOH$$

Hippuric acid

Stereochemical descriptors

All but structures (1), (3) and (14) in this section may appear as a single optical isomer. The stereo-designators are not R and S, but D and L. Their significance is as follows:

where the central carbon is the so-called 'alpha-carbon atom'.

If it is desired that the isomer be distinguished in the name, the D or L comes immediately before the acid name-stem, separated from it by a hyphen.

In the case of glycine and hippuric acid there is no such isomeric possibility unless the CH_2 group is substituted. D or L is then used with the same stereo-significance but the symbol comes at the start of the name or, in the case of salts and esters, in the anion name immediately before the name(s) of the group(s) attached to the chiral C-atom. For examples, see section 22.5.

22.3 AMINOSULPHONIC ACIDS

The only trivial name under this heading is:

$$H_2N-CH_2CH_2-SO_2-OH \qquad \text{Taurine}$$

It is retained as a PFS for its salts and esters, substituted derivatives other than alkyl groups on the carbon-chain and for salts, but not for halides or amide-derivatives of the parent acid.

22.4 DERIVATIVES OF THE TRIVIALLY NAMED ACIDS OF SECTIONS 22.2.1–22.2.5

22.4.1 Salts and esters

Salts are named by citing the metal or cation, then (after a space) the anionic form of the trivially named acid chosen, as appropriate, from sections 22.2.1–22.2.5. In the case of acids ending in '-ine' this is replaced by '-inate'. Otherwise 'ate' replaces 'ic acid'.

Example (1)

$$HO—CH_2—COOK$$

Potassium glycolate

Example (2)

$$[CH_3)_3NH]^+[H_2N — C(=O)CH_2CH_2COO]^-$$

Trimethylammonium succinamate

Example (3)

$$Mg^{2+}[H_2N — CH_2CH_2SO_2O^-]_2$$

Magnesium ditaurinate

Esters are named by citing the radical R (see Appendix C), then, after a space, the anion-form of the appropriate acid.

Example (4)

$$CH_3C(=O)—CH_2COOCH_2CH_3$$

Ethyl acetoacetate

Example (5)

$$CH_3—CH(NH_2)—COOC(CH_3)_3$$

tert-Butyl alaninate

Example (6)

$$HO—CH_2—COO—CH_2CH_2CH_2—O—C(=O)—CH_2OH$$

Trimethylene diglycolate

See also section 22.5.

22.4.2 Acyl halides

These are named by replacing the ending 'ic acid' or 'ine' in the appropriate acid by 'oyl halide' or 'yl halide', respectively.

Example (1)

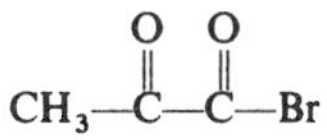

$$CH_3-\overset{\displaystyle O}{\overset{\|}{C}}-\overset{\displaystyle O}{\overset{\|}{C}}-Br$$

Pyruvoyl bromide

Example (2)

$$CH_3-CH(NH_2)-\overset{\displaystyle O}{\overset{\|}{C}}-I$$

Alanyl iodide

Exception

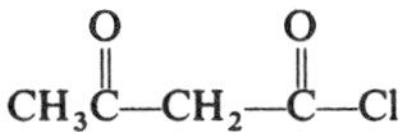

$$CH_3\overset{\displaystyle O}{\overset{\|}{C}}-CH_2-\overset{\displaystyle O}{\overset{\|}{C}}-Cl$$

Acetoacetyl chloride

22.4.3 Nitriles

These are regarded as derived from parent acids, the –COOH group being formally replaced by a –CN group. Accordingly, they are named by changing the ending 'ic acid' to 'onitrile'. In the case of acids ending in 'ine', the final 'e' is replaced by 'onitrile'.

Example (1)

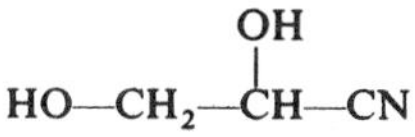

$$HO-CH_2-\overset{\displaystyle OH}{\overset{|}{CH}}-CN$$

Glyceronitrile

Example (2)

$$H_2N-CH_2-CN$$

Glycinonitrile

Example (3)

$$CH_3-CHOH-CN$$

Lactonitrile

22.4.4 Carboxylic amides

These are named by replacing the ending 'ic acid' or the final 'e' of the 'ine' ending of the appropriate acid by 'amide'.

Example (1)

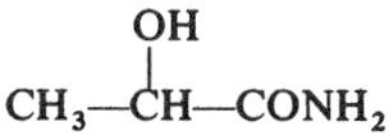

$$CH_3-\overset{\displaystyle OH}{\overset{|}{CH}}-CONH_2$$

Lactamide

Example (2)

$$CH_3—CH—CONH_2$$
$$|$$
$$NH_2$$

Alaninamide

N-phenyl derivatives are named as anilides (see section 25.14).

Example (3)

$$CH_2OH$$
$$|$$
$$H_2N—CH—CONH—\langle O \rangle—Br$$

4'-Bromoserinanilide

22.4.5   Aldehydes

Here the parent acid is regarded as having its –COOH group changed to –CHO and the name is formed by changing the ending of the appropriate acid-name replacing 'ic acid' by 'aldehyde'. In the case of the amino acids, 'al' replaces the final 'e'.

Example (1)

$$O$$
$$\|$$
$$CH_3C—CH_2—CHO$$

Acetoacetaldehyde

Example (2)

$$NH_2$$
$$|$$
$$H_2N—[CH_2]_4—CH—CHO$$

Lysinal

22.5   SUBSTITUTED DERIVATIVES

If the structure is based on an acid (or a derived function, i.e. –CON<,   —CHO, —COOM(R), —CN, —COX) named under section 22.1, the name is based on the appropriate stem-name in preference to any other mono-PG chain-structure. The same applies to identifiable derivatives of named compounds of section 22.2 and to those of taurine. In the event of a choice between trivially named base-structures, consult the yellow pages, section B.

Numbering

The PG carbon atom is numbered 1 and the carbon chain is numbered sequentially thereafter, as shown in the numbered examples of sections 22.1-22.4. Substituents on heteroatoms are given O- or N-locants as appropriate and subject to the stated restrictions of section 22.2.4. In cases where there is more than one heteroatom of the same kind, the site of substitution is distinguished by a superscript corresponding to the carbon atom to which it is attached e.g. N^2 or N^6 in (6) and N^2 or N^5 in (9) of section 22.2.5.

In the case of amino acids containing ring-systems, the chain has Greek letter locants whilst the rings have numerals (see the structures (3) and (4) of section 22.1 and (12) and (13) of section 22.2.5. Histidine is exceptional). The substituent radicals are cited in the name in alphabetical order before the name of the parent compound, formed as described above.

Example (1)

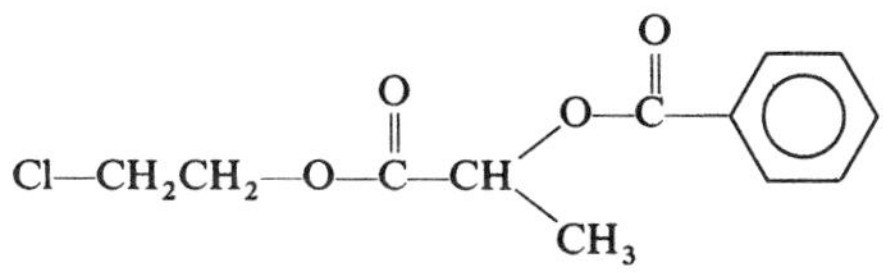

2-Chloroethyl *O*-benzoyl-lactate

Example (2)

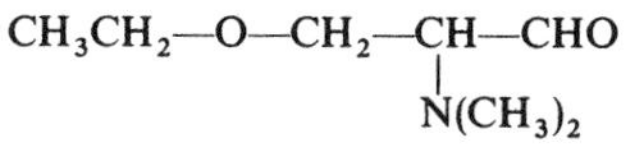

$$CH_3CH_2-O-CH_2-CH-CHO$$
$$N(CH_3)_2$$

O-Ethyl-*N*,*N*-dimethylserinal

Example (3)

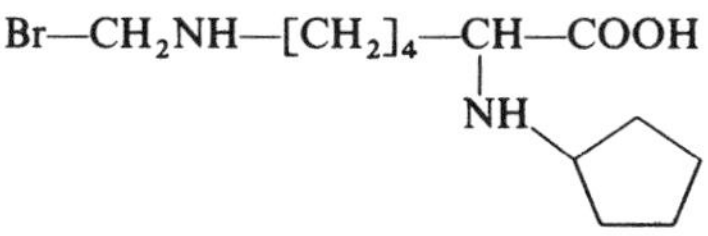

$$Br-CH_2NH-[CH_2]_4-CH-COOH$$
$$NH$$

N^6-Bromomethyl-N^2-cyclopentyl-lysine

Example (4)

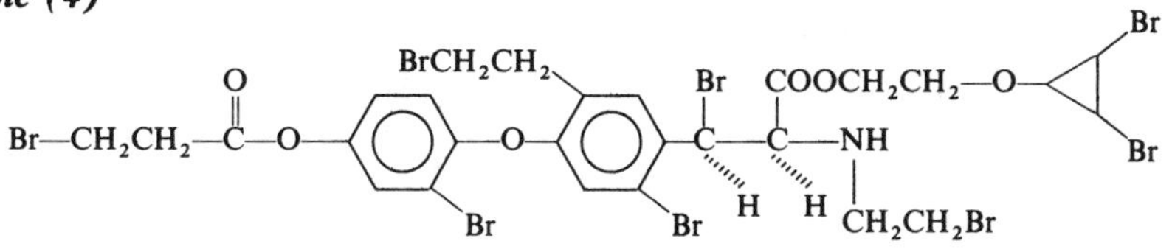

2-(2,3-Dibromocyclopropoxy)ethyl
(*R*)-β,2,2′-tribromo-*N*,5-bis(2-bromoethyl)-*O*-(3-bromopropionyl)-L-thyroninate

Example (5)

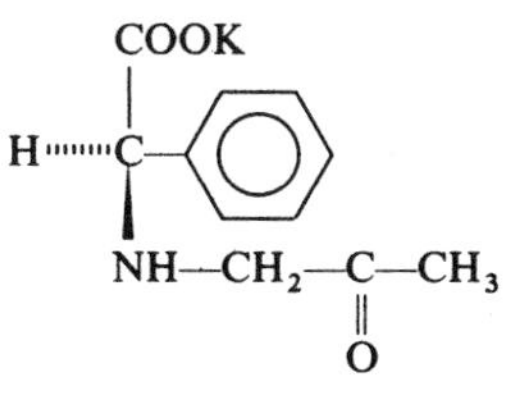

Potassium (*N*-acetonyl)–D-2-phenylglycinate

23 Vari-functional cyclic compounds†

23.0 INTRODUCTION

If your progress through the **flow-diagram** at the back of the book missed box (xxx), go directly to section 23.1 *et seq*.

If you arrived here via box (xxx), consult the yellow pages, in particular section C, and ascertain whether the PFS (see Appendix D-1.1) is a ring-system or a chain. If it is a chain, return to box (xxviii) of the **flow-diagram** and this time answer 'No' and proceed. If it is a ring-system, read on.

23.1 THE PG AND ONE OR MORE FGs COVERED BY A TRIVIAL NAME

Use of the following trivial name-stems is subject to two general provisos: (i) the ring must be a benzene ring†; if there is any hydrogenation of it, turn to section 24, and (ii) the benzene ring should not have directly attached to it the same subsidiary FG as that subsumed by the trivial name. If it does, the name is based on the appropriate name under section 24, e.g. 2,4-dimethoxybenzoic acid and not 4-methoxy-*o*-anisic acid; 3,4,5-triaminobenzenesulphonamide and not 3,5-diaminosulphanilamide.

To see if your PFS together with one or more FGs (defined in section 1.5) is covered by a trivial name, look at column 1 of Table 23.0 to identify the PG, and then at column 2 for the appropriate accompanying FG(s). Turn to the corresponding subsection. If no trivial name can be found there or, if your PG/FG combination is not covered, turn directly to section 24. When the PG is –COOH and the FGs include –COOM or –COOR, refer to sections 27.1.1 and 27.1.2 to see if both could be collected in a name after formal conversion of COOM or COOR to COOH. If so, see section 27.3.1; if not, see section 24.

---

† Except for section 23.1.4.

TABLE 23.0

Principal group (PG)	Functional group (FG)	Sub-section
—COOH	—CON< —CHO —OH —N< —OCH$_3$	23.1.1.1 23.1.1.2 23.1.1.3 23.1.1.4 23.1.1.5
—SO$_2$—OH	—N<	23.1.2
—COOM (salt) —COOR (ester)	—CON<; —CHO; —OH; —N<; —OCH$_3$	23.1.5.2[a]
—SO$_2$—OM (salt) —SO$_2$—OR (ester)	—N<	23.1.5.2[a]
—COX (acid halide)	—CON<; —CHO; —OH; —N<; —OCH$_3$	23.1.5.3[a]
—CON< (amide) —C(=N—)—N< (amidine)	—CHO; —OH; —N<; —OCH$_3$	23.1.5.4[a] 23.1.5.5[a]
—CN (nitrile)	—CHO; —OH; —N<; —OCH$_3$	23.1.5.6[a]
—CHO (aldehyde)	—OH; —N<; —OCH$_3$	23.1.5.7[a]
—CHO	—OH and —OCH$_3$	23.1.3
>C=O (ketone)	—OH	23.1.4

[a] The trivial stem-name of the corresponding acid is used for these functional derivatives.

23.1.1 Carboxylic acids

The following are used as the basis of names for derivatives with the numbering shown

23.1.1.1 Amic acids

Phthalamic acid   Isophthalamic acid   Terephthalamic acid

Where such trivially named amic acids have a benzene ring attached to the nitrogen atom of the amidic group, the name is modified to replace the ending '-amic acid' by 'anilic acid' [**Note** aniline ring-numbering is 'primed'].

Example

Phthalanilic acid

23.1.1.2 Aldehydic acids

Phthalaldehydic acid   Isophthalaldehydic acid   Terephthalaldehydic acid

23.1.1.3 Hydroxy acids

Salicylic acid
[used for the *o*-isomer only]

23.1.1.4 Amino acids

Anthranilic acid
[used for the *o*-isomer only]

23.1.1.5 Etheric acids

Anisic acid
[*p*-isomer shown]

23.1.2 Amino-sulphonic acids

The following are retained with the numbering shown:

Sulphanilic acid   Naphthionic acid
[used for the *p*-isomer only]   [used for 4-amino only]

23.1.3   Aldehydes

The following trivial name is retained with the numbering shown:

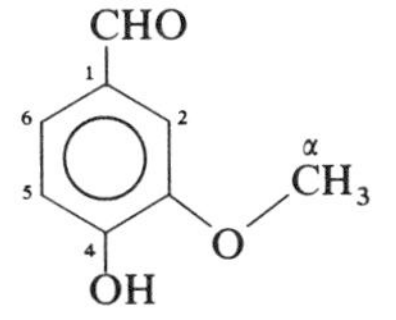

Vanillin

provided that there are no chain-alkyl substituents on the α-carbon atom. The only substitution allowed on the —OH group is an acyl group. If either proviso is not met, the PFS reverts to benzaldehyde (see section 24.17). For other trivially named aldehydes see section 23.1.5.7.

23.1.4   Acyloins

These have the general formula

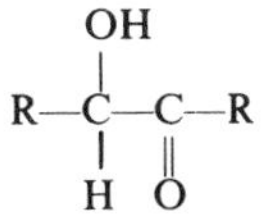

where R is an aryl or a heterocyclic radical.

They are named by removing the ending 'ic acid' from the corresponding acid R—COOH and replacing it by 'oin'.

Example (1)

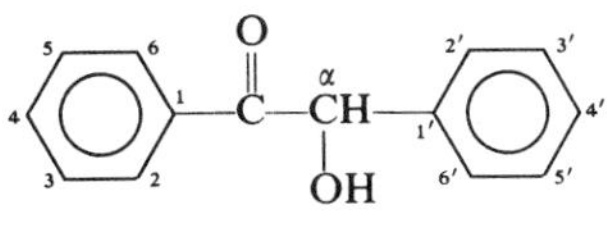

Benzoin

Example (2)

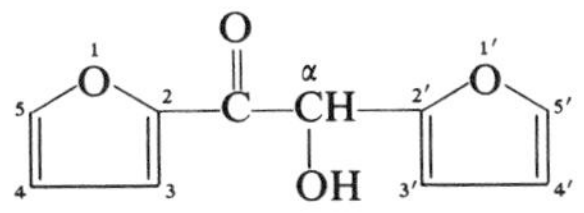

2,2'-Furoin

If the two R-groups are different, a name is derived under section 24.18.2.

23.1.5   Functional derivatives of acids of sections 23.1.1 and 23.1.2

23.1.5.1   Anions

These are named by converting the name-ending of the corresponding acid from 'ic acid' to '-ate', e.g. salicylate, *m*-anisate, sulphanilate.

23.1.5.2   Salts and esters

These are named by citing the name of the metal (or cation), or of the radical, as the case may be, and then, after a space, that of the appropriate anion from section 23.1.5.1.

Example (1)

COONa

CONH$_2$

Sodium isophthalamate

Example (2)

COOC$_2$H$_5$

NH$_2$

Ethyl anthranilate

For esters of the –OH group of salicylic acid see section 23.2.

23.1.5.3   Acyl halides

$$\overset{O}{\overset{\|}{-C}}-X \quad \text{or} \quad -SO_2X$$

These are named by converting the name-ending of the appropriate carboxylic acid from '-ic acid' to 'oyl halide', e.g. anthraniloyl fluoride.

In the case of sulphonic acids the corresponding conversion is from 'ic acid' to '-yl halide'.

23.1.5.4   Amides

$$\overset{O}{\overset{\|}{-C}}-N< \quad \text{or} \quad -SO_2-N<$$

[other than those of section 23.1.1.1, which are named under section 27.3 (q.v.) and those of section 23.1.1.2, which are named under section 24.12.1 (q.v.).]

These are named by converting the name-ending of the appropriate acid of sections 23.1.1.3 to 23.1.1.5 inclusive from '-ic acid' to '-amide'.

Example (1)

NH₂ appears on the ring; CONH₂ group attached.

Anthranilamide

Example (2)

SO₂NH₂ at position 1; NH₂ at position 4.

Sulphanilamide

Structures having a benzene ring (though not a biphenyl group) directly attached to the amide nitrogen atom are named by converting the ending 'amide' to '-anilide' [see example (3) of section 23.2].

23.1.5.5 *Amidines*

These can be considered as deriving from the corresponding carboxylic acid of sections 23.1.1.3–23.1.1.5.

They are named by replacing the ending '-ic acid' of the corresponding trivial acid name by '-amidine'.

Example (1)

OH group on ring; C—NH₂ with ‖NH below.

Salicylamidine

Amidines derived from acids of section 23.1.1.1 are named as amidino-substituted derivatives of benzamide (as under section 24.12.1). Those derived from acids of section 23.1.1.2 are named as formyl-substituted benzamidines (as under section 24.15). For N-substituted amidines, see under section 24.25.

Example (2)

N—CH₃, ‖, C—NHCH₃; C—NHCH₃ with ‖O.

2-(N^1,N^2-Dimethylamidino)-*N*-methylbenzamide

23.1.5.6   *Nitriles*

–CN

These are regarded as deriving from the corresponding –COOH acid and are named by replacing the name-ending 'ic acid' of the appropriate acid trivial name from sections 23.1.1.3–23.1.1.5 by '-onitrile', e.g. anthranilonitrile.

Nitriles of acids of section 23.1.1.1 are named as cyano-substituted benzamide (as under section 24.12.1); those of acids of section 23.1.1.2 are named as formyl-substituted benzonitrile (as under section 24.16).

23.1.5.7   *Aldehydes*

–CHO

This section considers the aldehydes derived from the acids of sections 23.1.1.3–23.1.1.5. [Those derived from section 23.1.1.1 are named as formyl-substituted benzamide (as under 24.12.1) and those from section 23.1.1.2 are named under section 27.3.1 (q.v.).]

The ending of the corresponding acid name is converted from 'ic acid' to '-aldehyde', e.g. salicylaldehyde; anthranilaldehyde.

23.2   SUBSTITUTED DERIVATIVES

If the structure is based on one of the trivially named compounds of section 23.1, that provides its name-ending, (subject to the general provisos at the start of section 23 and any special provisos under its individual sub-sections). Names for attached groups are cited before it in radical prefix-form (See Appendix C). Each is preceded by its appropriate locant (identical groups being collected by 'di', 'tri', etc. and by 'bis', 'tris' etc. if identically substituted).

Numbering

The numbering is shown on each structure and is preserved in substituted derivatives. Ring positions may offer a choice, in which case lower numbering for the substituent-set at the first point of difference is preferred, after fixed numbering considerations (see Appendix D) have first been satisfied.

N and *O* are 'lower' than Greek letters, which are lower than numerals. 'Unprimed' locants are lower than 'primed' locants of the same value. In the case of N-substituted derivatives the N-locant suffices if there is no other nitrogen atom. When there is, they are distinguished by means of a superscript appropriate to their position of attachment.

In the case of N-substituted amidines, those groups on the —N$<$ atom are preceded by the locant N^1; those on the =N— atom by the locant N^2.

In the case of substitution of the –OH group of a trivially named hydroxy compound, e.g. salicylic acid or an acyloin, the trivial name is retained only for acyl substituents (see Appendix C). For others the R—O— group is named as a composite radical prefix, e.g. phenoxy, 3-chloroquinoxalin-2-yloxy, and the PFS (see Appendix D-1.1) is accordingly named as though the R—O— group were absent.

Thus

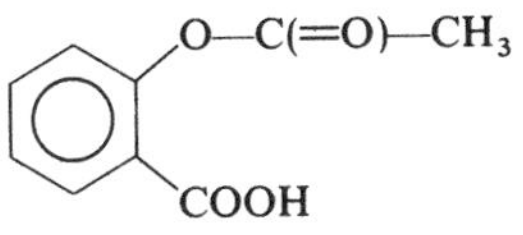

is named *O*-acetylsalicylic acid

but

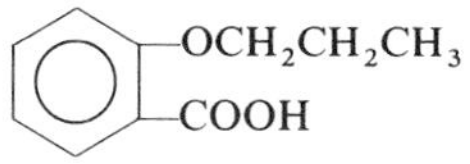

is named 2-propoxybenzoic acid

Example (1)

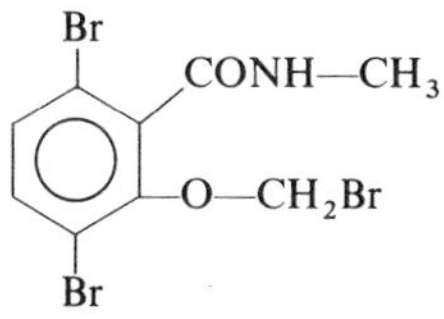

α,3,6-Tribromo-*N*-methyl
o-anisamide

Example (2)

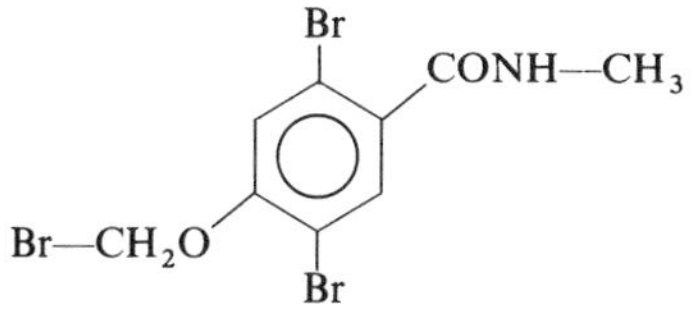

α,2,5-Tribromo-*N*-methyl
p-anisamide

[In the *p*-case the methoxy-site is 4 and the rest of the ring offers a numbering choice: 2,5- is lower than 3,6-.]

Example (3)

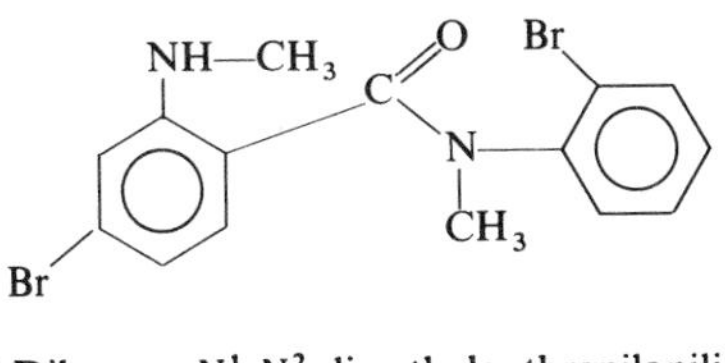

2′,4-Dibromo-*N*¹,*N*²-dimethylanthranilanilide

Example (4)

N^1,3-Dimethylsulphanilamide

24 Mono-PG ring-systems

24.0 PREAMBLE

If you reached box 24 of the **flow-diagram** at the back of the book without going via box (xxx) (whether or not redirected here from section 23) turn to the Table of contents and refer to the sub-section in section 24 appropriate to the PG (e.g. amides, ketones, phenols).

If you arrived here via box (xxx), consult the yellow pages to determine whether your PFS† is a chain- or a ring-system. If it is a chain, turn to section 25; if it is a ring system, refer to the Table of contents and then to the sub-section in section 24 appropriate to the PG [see also section 24.25].

The PG may be attached directly to a ring, e.g. –COOH, –CN, –CHO, or form part of a ring, e.g. lactone, imide. Some can occur in either situation, e.g. ketone, anhydride. This point has naming consequences and these are dealt with under the individual sub-sections as they occur.

24.1 CARBOXYLIC ACIDS

24.1.1 Trivially named

The following trivially named structures are used as a basis for naming substituted

† see Appendix D-1.1.

derivatives:

(1) Benzoic acid†

(2) Toluic acid
[p–shown]

(3) 1-Naphthoic acid

(4) 2-Naphthoic acid

(5) Furoic acid
[3- shown]

(6) Thenoic acid
[2- shown]

(7) Proline   (8) Nicotinic acid   (9) Isonicotinic acid

NB. The 'benzoic' and 'toluic' stems cannot be used when the ring has been hydrogenated to any degree (e.g. see example (3) of section 24.1.2).

If there is an alkyl group directly attached to the α-carbon of toluic acid, the name is based instead on benzoic acid. The same applies if there are any more CH_3-groups directly attached to the ring. If another phenyl group is directly attached to the ring, the name is based on 'biphenyl-x-carboxylic acid', where x is the appropriate locant.

The same considerations apply to the use of these same trivial name-stems for their related functional derivatives: salts, esters, amides, aldehydes, nitriles.

24.1.2 Systematically named

Acids other than the trivially named acids of section 24.1.1 are named by citing the appropriate ring-system from sections 6,7,9 or 11–20, then the locant (bounded by hyphens) for the –COOH group, and lastly the ending 'carboxylic acid'.

For monocycles of sections 6 and 7 the locant is 1 but it is omitted. In the latter case the double bond-set receives lowest locant(s) after that. For the ring-systems of sections 9 and 11–20, the locant is the lowest allowed by fixed numbering considerations.

† Biphenyl is exceptional (see section 24.1.2, example (2)).

Example (1)

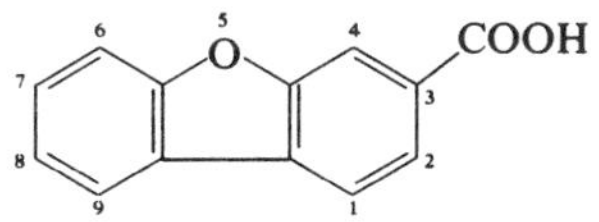

Dibenzofuran-3-carboxylic acid [not -7-]

Example (2)

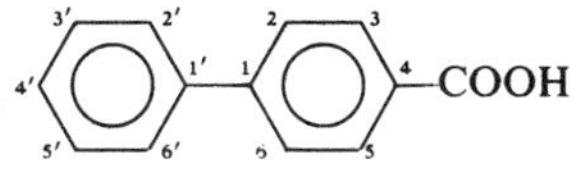

Biphenyl-4-carboxylic acid [not 4-phenylbenzoic acid]
Biphenyl derivatives are named as such (see section 12.2).

Example (3)

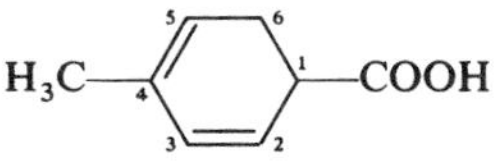

4-Methylcyclohexa-2,4-dienecarboxylic acid

24.2 PEROXY ACIDS

The structure 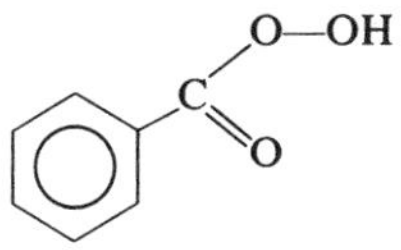 is called perbenzoic acid.

All others are named by adding 'peroxycarboxylic acid', preceded by its locant (flanked by hyphens), to the name of the appropriate ring-system. For assignment of locants, the final para of section 24.1.2 applies.

Example

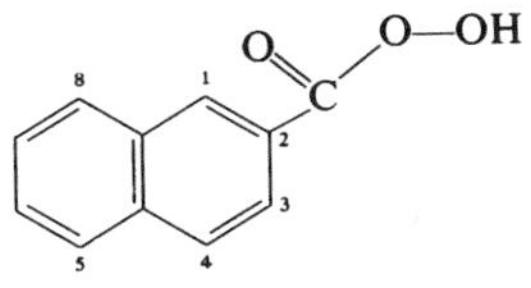

Naphthalene-2-peroxycarboxylic acid

24.3 CYCLIC ANHYDRIDES

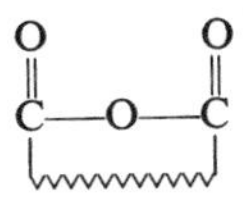

Turn to section 27.3.2.1.

24.4 THIOIC AND DITHIOIC ACIDS

When the group —C(=O)—SH is attached to benzene or naphthalene, the name is based on thiobenzoic acid or thio-1(or -2)-naphthoic acid, respectively.

In the case of the group —C(=S)—SH, the name is based on dithiobenzoic acid or dithio-1(or -2)-naphthoic acid, respectively.

For other acids, the name is formed from the name of the cyclic system taken, as appropriate, from sections 6, 7, 9 or 11–20, followed by the ending 'carbothioic acid' for the group –COSH, which is realistically:

$$C \overset{O}{\underset{S}{\diamondsuit}} \cdot H$$

but can be fixed in ester-formation as C(=O)—SR or C(=S)—OR (see section 24.8).

For the group –CSSH the ending is 'carbodithioic acid'.

For assignment of locants the final para of section 24.1.2 applies.

24.5 SULPHONIC ACIDS

—SO$_2$—OH

The name of the appropriate ring or ring-system is followed by the locant (flanked by hyphens) and then 'sulphonic acid'. For assignment of locants the final para of section 24.1.2 applies.

For halides and amides of sulphonic acids (–SO$_2$X; –SO$_2$NH$_2$), see sections 24.10 and 24.12.2, respectively.

24.6 ANIONS

These are formed from the name of the parent acid by changing the ending '-ic acid' to '-ate', e.g. benzoate, isonicotinate, thenoate, naphthalene-2-sulphonate, furan-3-carbodithioate. For proline the anion is prolinate.

24.7 SALTS

–COOM

The name begins with that of the metal or the organic cation then, after a space, the name of the anion, formed by conversion of the name for the appropriate acid cited in section 24.1 or 24.2 to its anionic form, e.g. sodium benzoate, or tetraphenylammonium cyclopropanecarboxylate.

24.8 ESTERS

–COOR

The naming procedure is exactly analogous to that of section 24.7 except that the metal (or cation) name is replaced by that of the appropriate radical-group in prefix-form (see Appendix C).

Example (1)

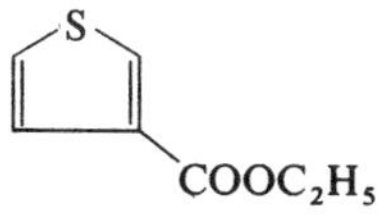

Ethyl 3-thenoate

For esters of monothioic acids the radical is preceded by the appropriate element symbol *O-* or *S-*.

Example (2)

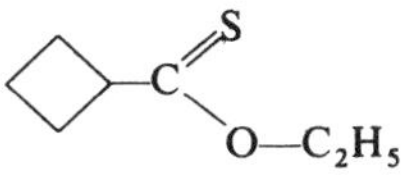

O-Ethyl cyclobutanethioate

24.9   LACTONES AND LACTAMS

24.9.1   Lactones

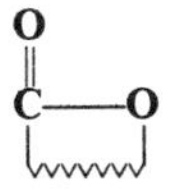

These may be thought of as internal esters of hydroxyacids, formed by elimination of water. The following names are retained as parents for naming derivatives:

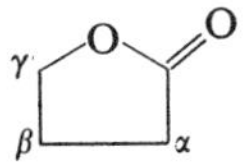

(1) γ-Butyrolactone

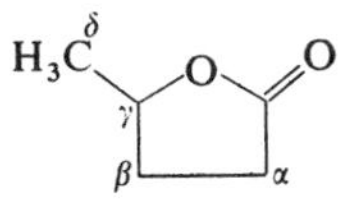

(2) γ-Valerolactone

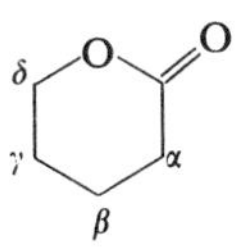

(3) δ-Valerolactone

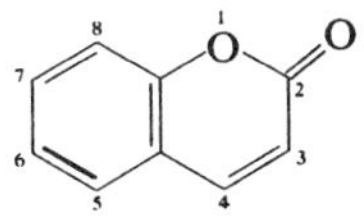

(4) Coumarin

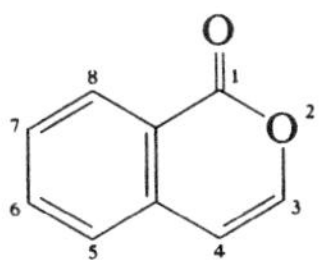

(5) Isocoumarin

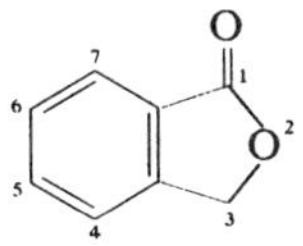

(6) Phthalide

Structures (1)–(3) are for saturated forms only [cf. example (8) in this section].

Other single-ring lactones are named by cutting the ring between the ring oxygen atom and the —C(=O) group, then replacing the ring oxygen atom by a hydrogen atom and the >C=O by a –CH$_3$ group. The resulting hydrocarbon name from section 5.1 is cited (without its final 'e'), followed by the locant (flanked by hyphens) of the carbon atom which had been connected directly to the ring oxygen atom on the other side from the >C=O group, then 'olide'.

For this purpose the >C=O position is numbered 1 and numbering proceeds sequentially around the lactone ring in the direction leading away from the ring oxygen atom. This numbering is retained for name-construction in the transformed open chain.

Example (7)

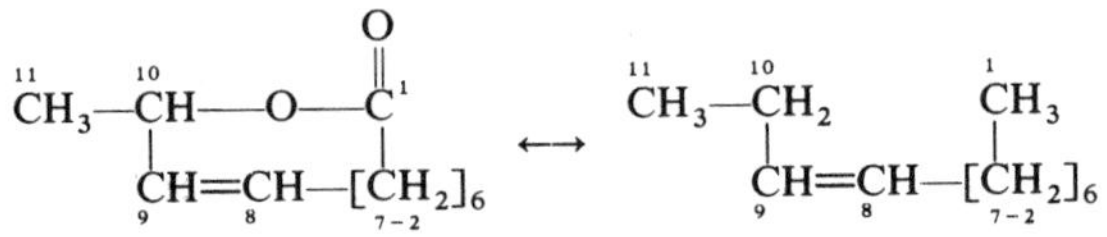

Undec-8-en-10-olide

Example (8)

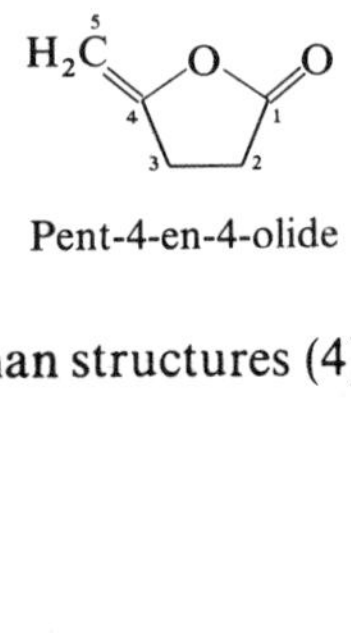

Pent-4-en-4-olide

Bi- and polycyclic lactones other than structures (4)–(6) are named as cyclic ketones as under section 24.18.3.

24.9.2   Lactams

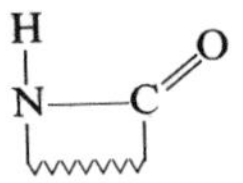

These are named as cyclic ketones under section 24.18.3 (q.v.).

24.10   ACID HALIDES

–COX   or   –SO$_2$X

The appropriate name of the parent acid R—COOH or R—SO$_2$OH is transformed to signify its replacement by R—COX or R—SO$_2$X, respectively, where R is a cyclic system:

-oic acid	becomes	-oyl halide
-ine	becomes	-yl halide
-carboxylic acid	becomes	-carbonyl halide
-sulphonic acid	becomes	-sulphonyl halide

24.11   UREA DERIVATIVES

$$(R^1 \text{ or } H)(R^2 \text{ or } H)N-\overset{\overset{\textstyle O}{\|}}{C}-N(R^3 \text{ or } H)(R^4 \text{ or } H)$$

$(R^1-R^4 \text{ may be the same or different.})$

Each ring-system is expressed as a radical prefix (see Appendix C). The nitrogen atom carrying the greater number of substituent groups is numbered 1 and the other 3. If each carries an equal number of groups whether cyclic or not, the one cited first in the name (alphabetically) is regarded as being attached to position 1.

When the urea bears a single substituent, the locant for it is omitted.

When the urea grouping forms part of a ring, thus:

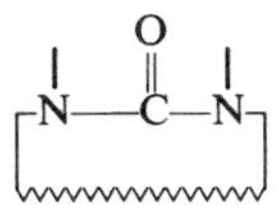

the name is formed as under section 24.18.3 (q.v.).

Example (1)

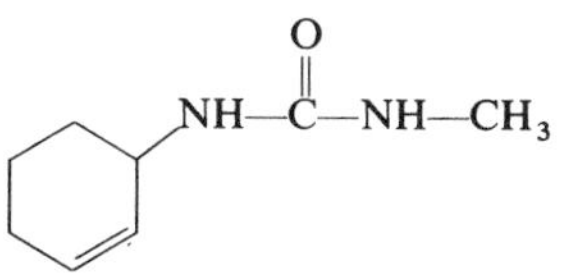

1-Cyclohex-2-enyl-3-methylurea
[order of citation: c,m]

Example (2)

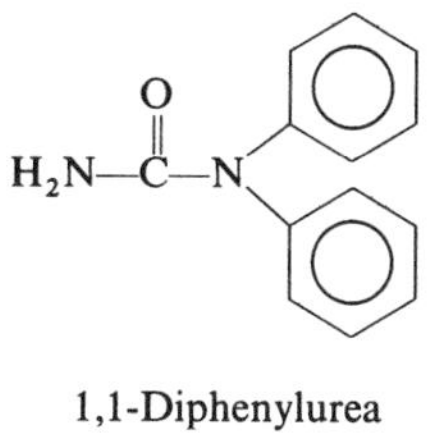

1,1-Diphenylurea

Example (3)

$$\text{H}_2\text{C} \diagdown \atop \text{H}_2\text{C} \diagup \text{CH}-\text{NH}-\overset{\overset{\text{O}}{\|}}{\text{C}}-\text{NH}_2$$

Cyclopropylurea

24.12 AMIDES

24.12.1 Carboxamides

$$-\text{CONH}_2$$

The name-ending of the corresponding trivially named carboxylic acid is converted from '-oic acid' to 'amide', e.g. 2-naphthoic acid → 2-naphthamide.

The amide-name from proline is 'prolinamide'.

In the case of systematically named acids, the ending '-ylic acid' is converted to 'amide', e.g. cyclobutanecarboxylic acid → cyclobutanecarboxamide.

24.12.2 Sulphonamides

For sulphonic acid amides $\text{R}-\text{SO}_2\text{N}<$ the conversion is '-ic acid' to 'amide', e.g. benzenesulphonic acid → benzenesulphonamide.

24.12.3 Carbothioic amides

$$-\overset{\overset{\text{S}}{\|}}{\text{C}}-\text{N}<$$

The name is formed by modifying that of the corresponding acid. Benzoic acid and naphthoic acid become thiobenzamide and thionaphthamide, respectively.

Systematically named acids have their endings changed from '-ic acid' to '-amide', e.g. cyclobutanecarbothioamide.

24.13 HYDRAZIDES

$$-\overset{\overset{\text{O}}{\|}}{\text{C}}-\text{NH}-\text{NH}_2 \; ; \qquad -\overset{\overset{\text{O}}{\uparrow}}{\underset{\text{O}}{\text{S}}}-\text{NH}-\text{NH}_2$$

Replacement of the $-\text{NH}-\text{N}<$ group by $-\text{OH}$ gives the corresponding acid. In the case of benzoic and naphthoic acids, the hydrazide name is formed by conversion of its name-ending from '-ic acid' to '-hydrazide', e.g. 2-naphthohydrazide.

For other acids the appropriate conversion is from 'carboxylic acid' to '-carbohydrazide' and from 'sulphonic acid' to 'sulphonohydrazide'.

24.14 IMIDES

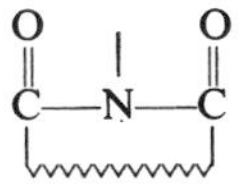

$$\underset{\displaystyle \overset{\displaystyle \|}{\text{C}}}{\overset{\displaystyle \text{O}}{}}\!-\!\underset{\displaystyle \|}{\text{N}}\!-\!\underset{\displaystyle \overset{\displaystyle \|}{\text{C}}}{\overset{\displaystyle \text{O}}{}}$$

See section 27.3.2.3.

24.15 AMIDINES

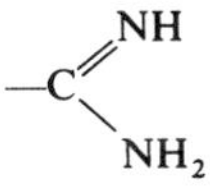

$$-\text{C}\underset{\displaystyle \text{NH}_2}{\overset{\displaystyle \text{NH}}{\Big\langle}}$$

These are considered to be derived from the acids of section 24.1 by replacement of the –COOH group by the above group.

The appropriate acid name is transformed as follows:

-ic acid or -oic acid	becomes	-amidine
-carboxylic acid	becomes	-carboxamidine

Example (1)

$$C_6H_5 . C(=NH)\!-\!NH_2$$

Benzamidine

Example (2)

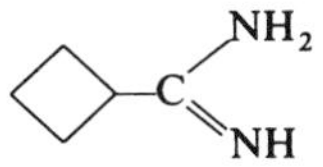

Cyclobutanecarboxamidine

24.16 NITRILES

–CN

In the case of the trivially named acids of section 24.1.1 the derived nitrile, (in which the –COOH is considered to be modified to –CN), is named by changing the ending 'ic acid' to 'onitrile'. Proline is modified to prolinonitrile.

Example (1)

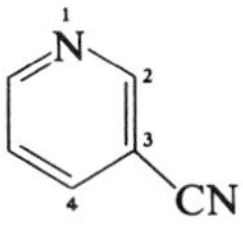

Nicotinonitrile

Other nitriles are named by citing the appropriate name for the ring or ring-system and then the locant (flanked by hyphens), then 'carbonitrile'. For assignment of locants the final para of section 24.1.2 applies.

Example (2)

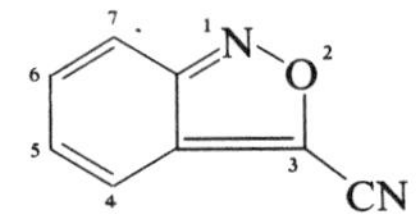

2,1-Benzoxazole-3-carbonitrile

24.17 Aldehydes

—CHO

Those aldehydes derived from the trivially named acids are named by changing the ending: 'ic acid' or 'oic acid' to 'aldehyde'.

Exception (1)

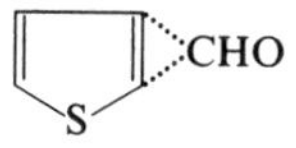

Thiophene-2(or 3)-carbaldehyde

Exception (2)

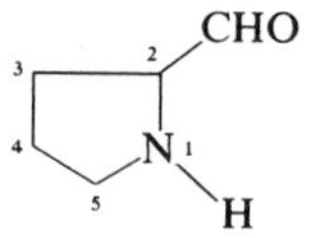

Prolinal

Others are named by adding to the name of the ring system the appropriate locant (if any) and 'carbaldehyde'. The final para of section 24.1.2 applies.

24.18   KETONES

$$\gtrdot C{=}O$$

If you reached section 24 via box (xxx), turn to section 21.3.3; if not, read on.

24.18.1   The $\gtrdot C{=}O$ group attached directly to two separate ring-systems

(a) At C-sites in both cases: turn to section 21.3.1.

(b) One at a heteroatom ring-site. To construct the name, first remove this heteroatom and its associated ring-system, so generating a carbonyl radical (see Appendix C-3). The name then consists of the appropriate radical prefix, whether trivial or systematic, preceded by the locant for its point of attachment and a hyphen, followed by the name of the ring system that has been detached.

Example (1)

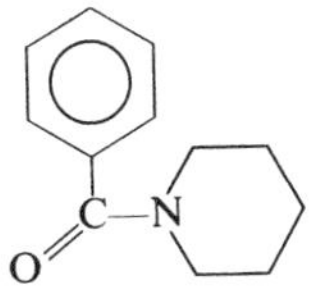

1-Benzoylpiperidine

Example (2)

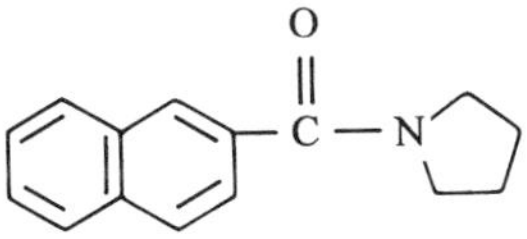

1-(2-Naphthoyl)pyrrolidine

Example (3)

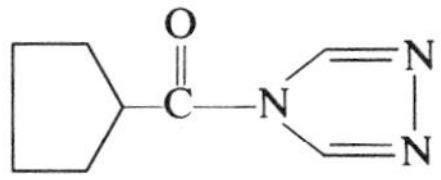

4-(Cyclopentylcarbonyl)-4*H*-1,2,4-triazine

(c) Both sides joined to the $>$C$=$O via a heteroatom. The name is based on that of the senior ring-system and the rest of the structure is cited in the name in the form of an appropriate radical prefix (ending in 'carbonyl'—see Appendix C), with appropriate locants.

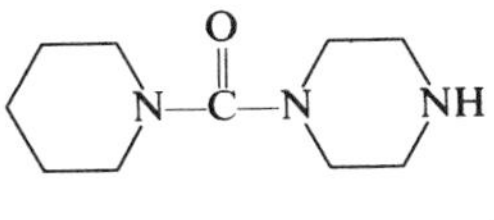

1-(1-Piperidylcarbonyl)piperazine

24.18.2   The $>$C$=$O group attached also to a chain

If the ring-system is benzene or naphthalene, then the name ends in 'phenone' or 'naphthone', respectively. The chain is expressed as the appropriate acyl radical (See Appendix C), but with the 'yl' ending changed to 'o'. Ring locants are 'primed'.

Example (1)

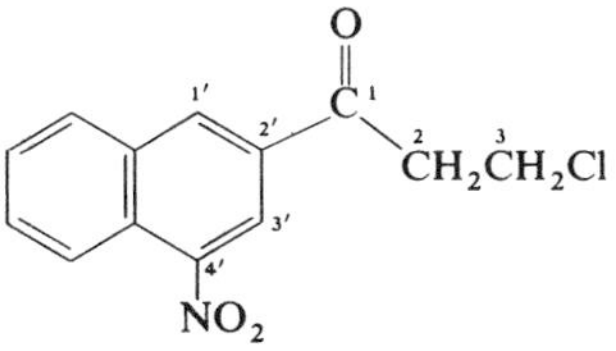

3-Chloro-4'-nitro-2'-propiononaphthone

Exception

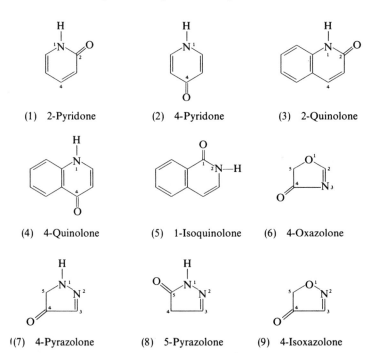

Propiophenone

For other ring-systems, the chain leading up to and including the —C(=O) group is named as in section 5. It is cited, preceded by the name of the ring-system in its appropriate radical prefix form (see Appendix C3 & C4) and followed by the ending '-1-one'.

In the case of methanone and ethanone the l-locant is omitted.

Example (2)

1-(2-Methyl-2*H*-chromen-2-yl)butan-l-one

24.18.3   The $>$C=O group occupying a ring position

The following set of contracted forms are used for the structures shown, and only those*. Any change in bonding or position of the =O group from that shown in each case results in the name being formed systematically instead.

(1)   2-Pyridone   (2)   4-Pyridone   (3)   2-Quinolone

(4)   4-Quinolone   (5)   1-Isoquinolone   (6)   4-Oxazolone

((7)   4-Pyrazolone   (8)   5-Pyrazolone   (9)   4-Isoxazolone

*This section covers heterocyclic keto-compounds in which the $>$C=O group is not flanked by C-atoms as required by the classical definition of "ketone". When such structures are not named as lactones of section 24.9.1 they are named by whichever is the more suitable method from this section.

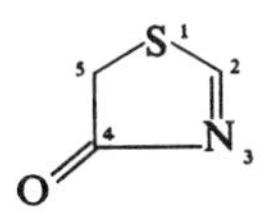

(10) 4-Thiazolone 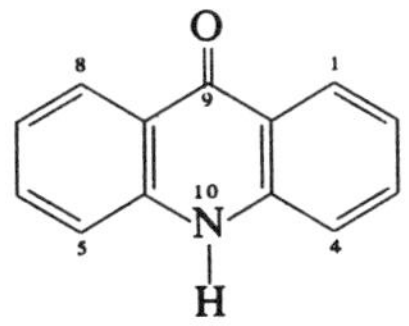   (11) 9-Acridone

If none of these structures is appropriate, the following methods are for general use:

Method A

The ring-system is named in the appropriate state of hydrogenation and 'one' added, preceded, where necessary, by its appropriate ring-locant.

Example (1)

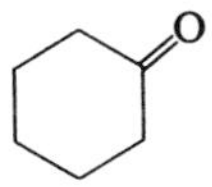

Cyclohexanone
[no locants needed]

Example (2)

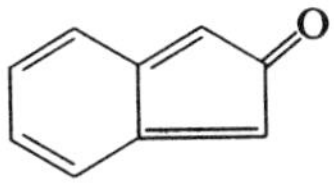

2H-Inden-2-one
['e' of indene elided before 'o']

Example (3)

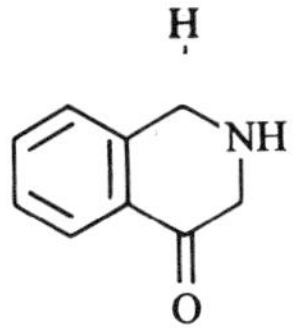

1,2,3,4-Tetrahydroisoquinolin-4-one

Method B

The ring-structure complete with its $>C=O$ group is drawn with the maximum possible number of ncdbs [see section 1.5] added. If this can be done with no spare $>CH_2$–positions and thus no choice of position for double bonds, then there is no need for indicated† hydrogen; if it cannot be done thus, then there is such a need.

---

† The significance of this term is explained in section 16.3.

Example (4)

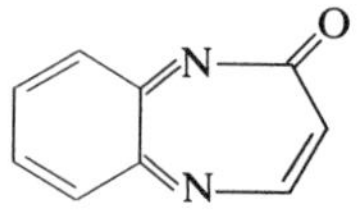

1,5-Benzodiazepin-2-one

Example (5)

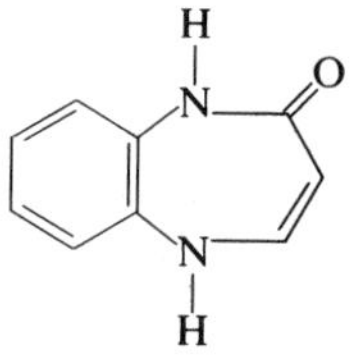

1,5-Dihydro-1,5-benzodiazepin-2-one

Example (6)

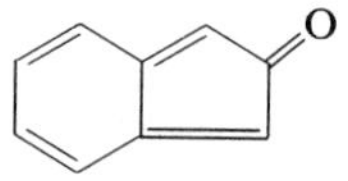

Inden-2-one

Example (7)

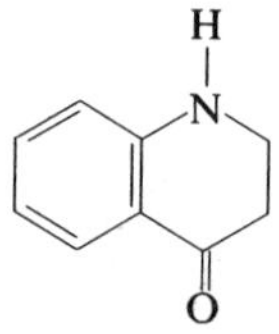

2,3-Dihydro-4-quinolone
[cf. structure (4) of section 24.18.3]

Example (8)

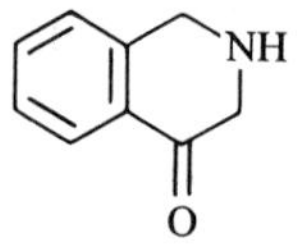

2,3-Dihydroisoquinolin-4(1*H*)-one

If a $>$C=O group is introduced into isoquinoline (section 15.1.1 structure 6) at position 4, the maximum number of ncdb's (see section 1.5) is reduced from 5 to 4. Of all the possible isomers so formed, that having the lowest locant for the H-atom so generated is, in this case, 1. The structure of example (8) is named as the 2,3-dihydro derivative of that isomer, rather than as the 1,2-dihydro derivative of the 3*H*-isomer (cf. the last sentence of section 16.3).

However, this method provides a possible exception to this rule when the site of the oxo-group claims lowest locant in the event of a choice e.g.

Example (9)

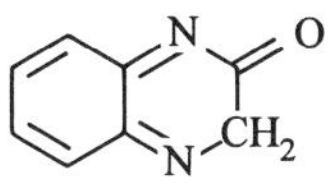

Quinoxaline-2(3*H*)-one

Method A is generally more suited to hydrogenated forms; Method B is suited to mancunide† or near-mancunide structures. Thus the name for structure (6) is preferred to that given under example (2), whereas the name for structure (3) is preferred to that given under example (8). As a guide, prefer (i) minimal use of hydro-prefixes and indicated hydrogen, and then (ii) one or the other rather than both.

24.19 THIOKETONES

$$>\!C\!=\!S$$

The '-one' endings of the ketone names of section 24.18 are replaced by 'thione'; any elided 'e's are now restored.

Example (1)

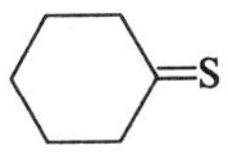

Cyclohexanethione

Exception

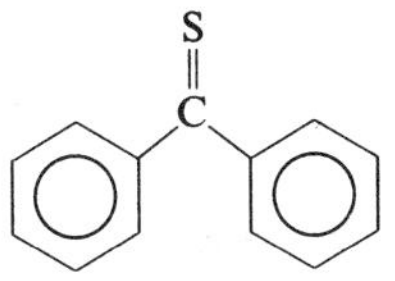

Thiobenzophenone

For other S-derivatives of phenones and naphthones the name of the side-chain is cited with the radical prefix for the ring-system preceding it and the ending '-l-thione'.

Example (2)

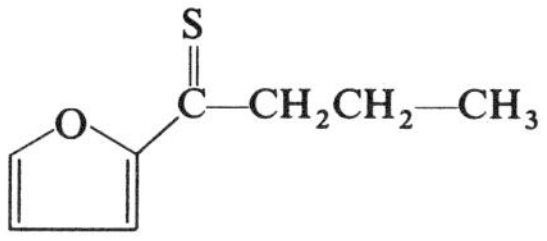

1-(2-Furyl)butane-l-thione

† Defined in section 1.5.

The $>$C$=$S group attached directly to two different ring-systems is given a three-word name of the type described in section 21.3.1, example (3).

24.20   OXIMES

$$=N-OH$$

See section 21.5.

24.21   ALCOHOLS AND THEIR METAL DERIVATIVES

24.21.1   Trivially named alcohols

The following trivial names are used where appropriate:

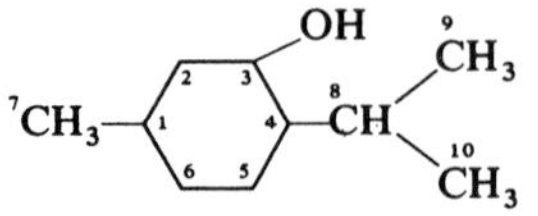

Menthol 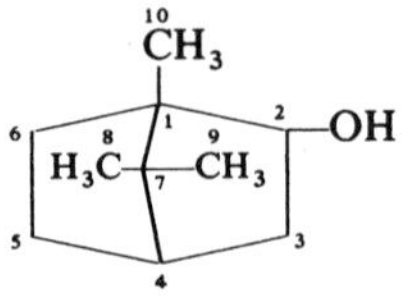   Borneol

24.21.2   Systematically named alcohols

Other alcohols are named by adding 'ol' to the name of the ring-system, these two parts of the name being separated by the hyphen-flanked locant, if any. For assignment of locants the final para of section 24.1.2 applies. Note that 'ol' implies an −OH group attached to a carbon atom; if it is attached instead to a heteroatom of a ring, the name is based on that of the heterocyclic system and the −OH group is expressed in the name as a hydroxy-prefix.

Example (1)

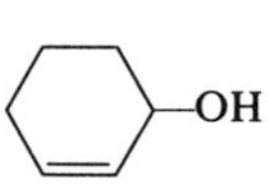

Cyclohex-2-enol

Example (2)

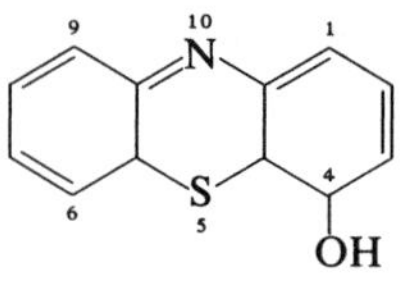

4,4a-Dihydro-5a*H*-phenothiazin-4-ol

[The name "phenothiazine" implies 6 double bonds. The lowest-numbered site for the indicated H compatible with that number is 5a. Example (2) is the 4,4a-dihydro derivative.]

Example (3)

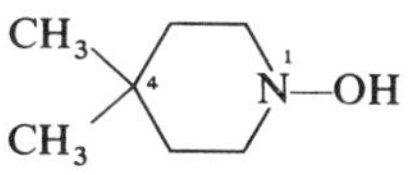

1-Hydroxy-4,4-dimethylpiperidine

24.21.3   Metal derivatives of alcohols

The name begins with that of the metal then, after a space, the radical prefix (see Appendix C) appropriate to the ring-system, then another space and the name 'oxide'.

Example (1)

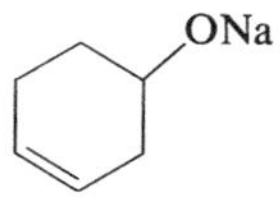

Sodium cyclohex-3-enyl oxide

Example (2)

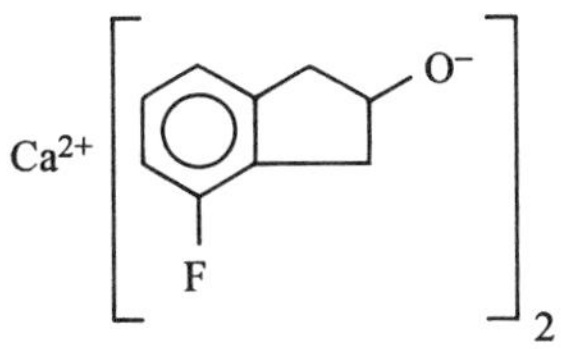

Calcium bis(4-fluoroindan-2-yl oxide)

Example (3)

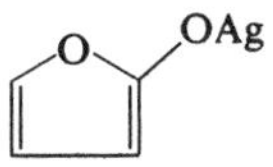

Silver 2-furyl oxide

24.22   PHENOLS AND THEIR METAL DERIVATIVES

24.22.1   Trivially named phenols

The following trivial names are preferred and used for naming substituted derivatives, subject to the provisos of section 24.25 (q.v.).

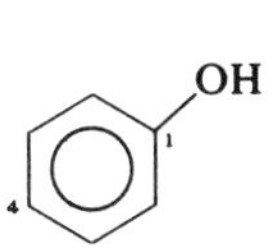

(1)   Phenol

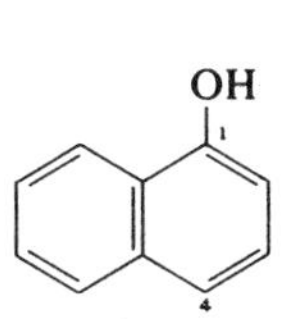

(2)   1-Naphthol

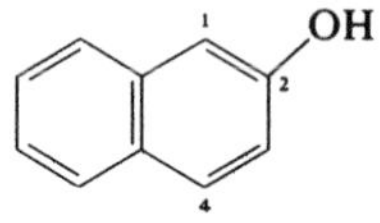

(3) 2-Naphthol

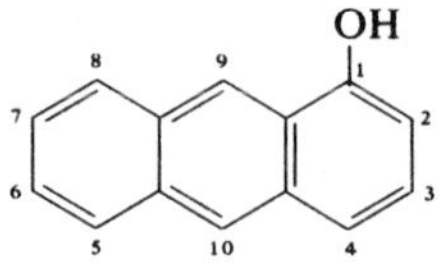

(4) Anthrol
[1- shown]

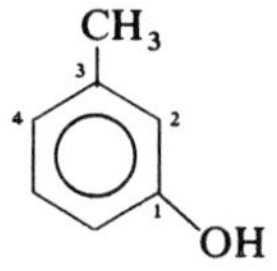

(5) Cresol
[m- shown]

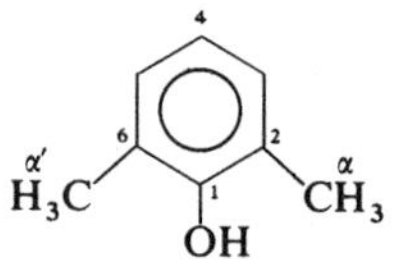

(6) Xylenol
[2,6- shown]

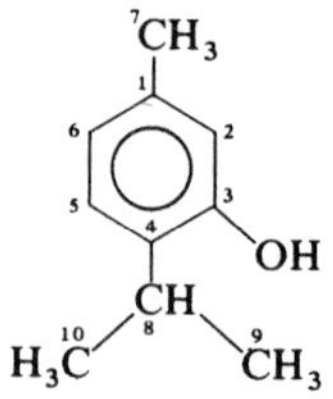

(7) Thymol

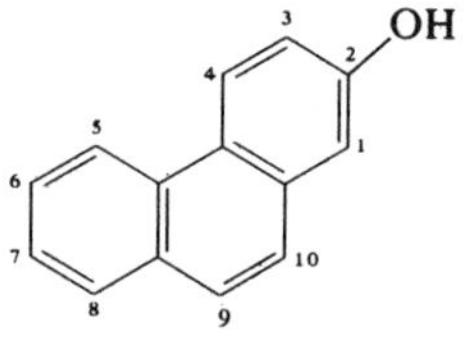

(8) Phenanthrol
[2- shown]

In this group, structures (2), (3), (4) and (8), being contracted forms, have their OH-locant before the stem and not before the '-ol' suffix. In the case of cresols and xylenols the locants in the name refer to the CH_3 groups, the −OH group always being at position 1.

24.22.2 Systematically named phenols

For other ring-systems, the locant and the '-ol' ending follow the appropriate ring-structure name from sections 12 or 18.

Example (9)

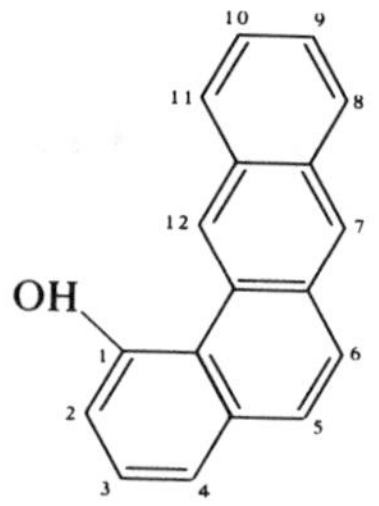

Benz[a]anthracen-1-ol

Example (10)

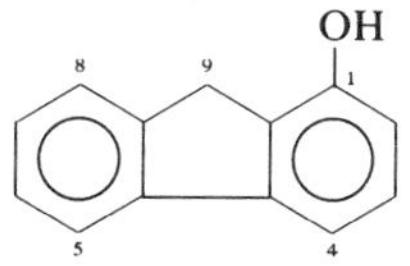

OH

Fluoren-1-ol
[not -8-]

24.22.3 Hydrogenated forms

For any hydrogenation of phenol, cresol, xylenol or thymol, see under section 24.21. However, other trivially named phenols keep their name-stem in all states of hydrogenation.

Example (1)

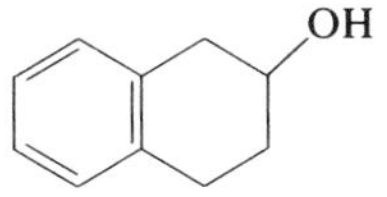

OH

1,2,3,4-Tetrahydro-2-naphthol

Example (2)

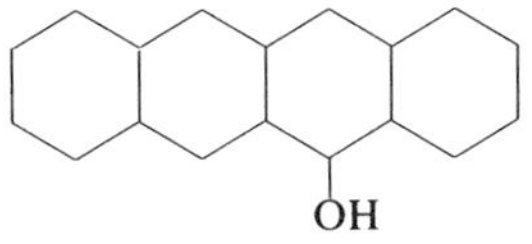

OH

Perhydronaphthacen-5-ol
[not -6-, nor -11-, nor -12-]

24.22.4 Metal derivatives of phenols

Those derivatives of the form: —OM

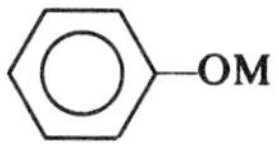

are named by citing the name of the metal, a space and then 'phenoxide'. For others, 'phenoxide' is replaced by the radical prefix appropriate to the ring-system, a space and then 'oxide'.

Example (1)

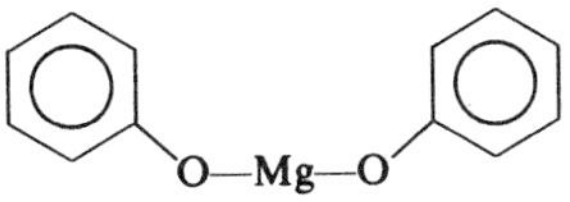

O—Mg—O

Magnesium diphenoxide

Example (2)

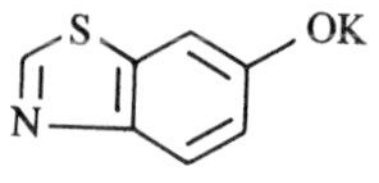

Potassium 1,3-benzothiazol-6-yl oxide

24.23 ESTERS OF INORGANIC ACIDS (OTHER THAN HX)

Names are constructed by citing the name of the cyclic radical, a space and then the name of the inorganic anion. These in turn are formed from the name of their parent acid by converting the endings '-ic acid' to '-ate' and '-ous acid' to '-ite', respectively.

Acid esters have 'hydrogen' inserted between the radical and the anion but joined only to the latter. Multiplicative prefixes are used as appropriate; 'mono' is invariably omitted.

Example (1)

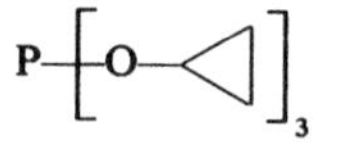

Tricyclopropyl phosphite

Example (2)

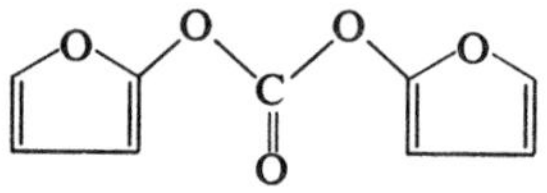

Di-2-furyl carbonate

Example (3)

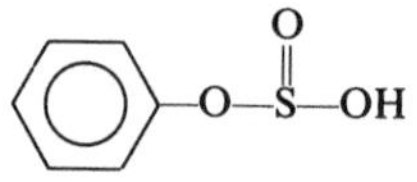

Phenyl hydrogensulphite

Example (4)

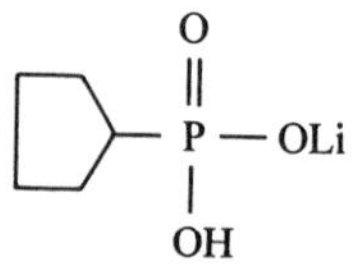

Lithium hydrogen cyclopentylphosphonate
[As the C-ring is joined *directly* to the P-atom, formally replacing a non-acidic H-atom, there is no space between "cyclopentyl" and "phosphonate". The resulting acid is organic and there is therefore a space following "hydrogen".

For salt/ester structures the order of citation is metal (or organic cation), radical, hydrogen (if any), anion—all separated by spaces except the 'hydrogen' and the anion.

Example (5)

Silver cyclobutyl hydrogenphosphate

24.24   THIOLS

The name is formed by citing first the name of the ring or ring-system, then a locant flanked by hyphens and finally 'thiol'. Benzene and cycloalkanes need no locant and in those cases it is omitted. In other cases, the locant numeral is as low as the constraints of ring-numbering allow (see section 24.1.2, final para).

Example (1)

Cyclopentanethiol

Example (2)

Quinoxaline-2-thiol

24.25   SUBSTITUTED DERIVATIVES

Special provisos, e.g. those under section 24.1.1 are observed in choosing the PFS.†.

The seniority of ring-systems, where there is a choice, is decided by criteria (A-2) *et seq.* yellow pages.

Numbering is as shown on the trivially named structures and that on others follows the general principles for numbering set out in Appendix D-2.1. Names for substituent groups (see Appendix C and Table 1 col.3) are cited in alphabetical order before that of the PFS. Each such group is preceded by its locant and a hyphen. The PG† is cited last, any final 'e' being elided before a vowel (e.g. as in '-ol').

† See section 1.5 & Appendix D-1.1.

Example (1)

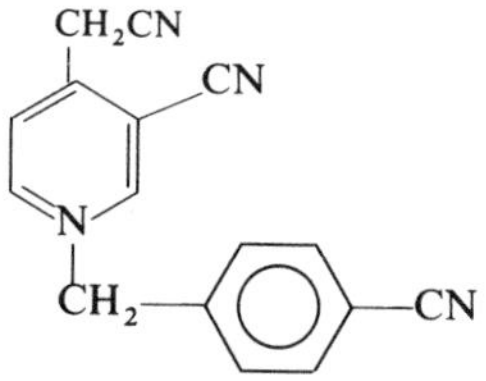

1-(4-Cyanobenzyl)-4-(cyanomethyl)nicotinonitrile
[Citation order: cyanob, cyanom. PFS decided by Yellow pages A(3) and C(3).]

For structures (1)–(3) of section 24.9.1, attached groups are given the appropriate Greek letter locants.

Example (2)

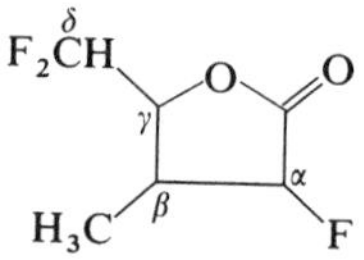

α,δ,δ-Trifluoro-β-methyl-γ-valerolactone

For N-substituted amides, N-locants are usually free from ambiguity when cited before the name of the parent amide.

However, when this could lead to ambiguity, the N-locant, followed by the appropriate prefix(es) is cited immediately before 'amide' inside parentheses.

Example (3)

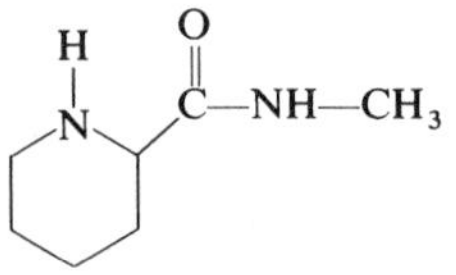

Piperidine-2-(*N*-methylcarboxamide)
[not *N*-methylpiperidine-2-carboxamide]

For substituted hydrazides the locant N is used for the nitrogen atom next to the $>C=O$ group and N' for the more remote nitrogen atom.

Example (4)

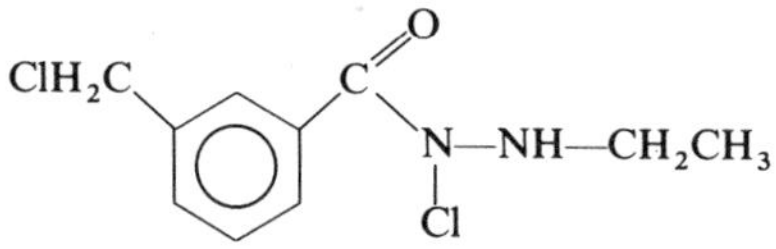

N,α-Dichloro-*N'*-ethyl-*m*-toluohydrazide

For N-substituted amidines, groups replacing the $=NH$ group hydrogen atom are preceded in the name by the locant N^2; those replacing a hydrogen atom of the NH_2 group by the locant N^1.

Example (5)

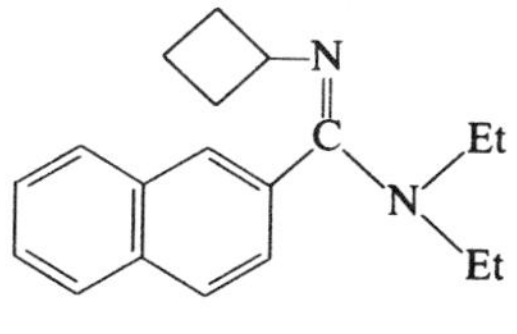

N^2-Cyclobutyl-N^1,N^1-diethyl-2-naphthamidine

For derivatives of benzophenone and thiobenzophenone the ring bearing the fewer substituents has its locants primed.

Example (6)

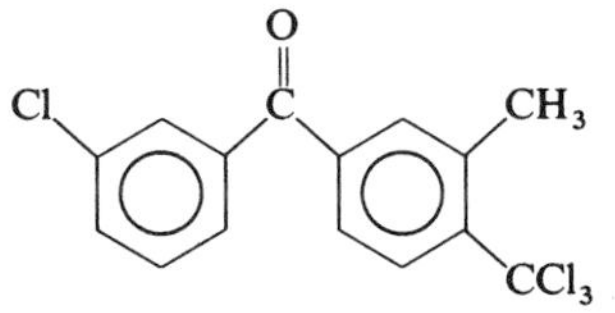

3'-Chloro-3-methyl-4-(trichloromethyl)benzophenone

Example (7)

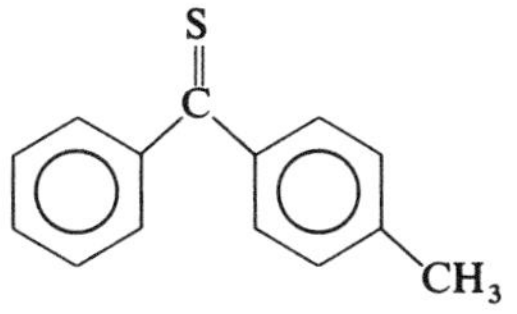

4-Methyl(thiobenzophenone)

Example (8)

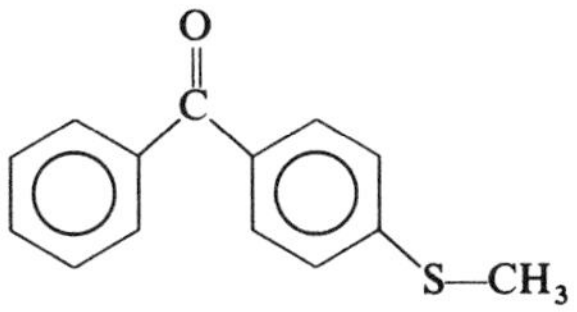

4-(Methylthio)benzophenone

For examples (7) and (8) note how ambiguity due to the two meanings of 'thio' is avoided by appropriate use of parentheses.

Menthol, borneol and thymol are not used as PFS (see Appendix D-1.1) when alkyl-substituted. In such cases the name is based on cyclohexanol, bicyclo[2.2.1]heptan-2-ol, and phenol, respectively.

When the PG is –COOH and –COOM (or –COOR) is also present but, after formal conversion from –COOM (or –COOR) to –COOH, the two –COOH groups are not collectable* in a named PFS*, the PFS is the ring-system bearing the –COOH and the name is formed according to Section 24.1, the –COOM or –COOR group being cited in prefix-form (See Table 1 column 3).

*"Collected" and "PFS" are defined in section 1.5.

Example (9)

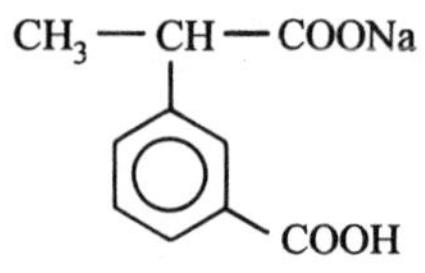

$$CH_3 - CH - COONa$$

COOH

3-[1-(Sodio-oxycarbonyl)ethyl]benzoic acid

Example (10)

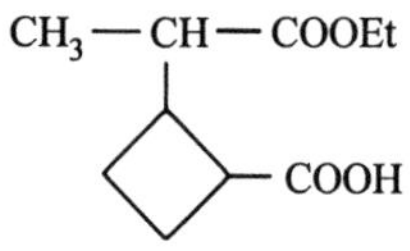

$$CH_3 - CH - COOEt$$

COOH

2-[-(Ethoxycarbonyl)ethyl]cyclobutanecarboxylic acid

25 Mono-PG chains

The presence of one or more rings in the structure does not prevent appropriate consideration under this section so long as the PG (defined in section 1.5) (a) is not directly attached to a ring, and (b) does not form part of a ring.

See section 4 before proceeding.

If you reached box 25 in your progress through the **flow-diagram** at the back of the book via box (xxx), whether or not redirected here from section 22, consult the yellow pages, section B, in order to determine the PFS (see Appendix D-1.1). If not, read on.

25.1 CARBOXYLIC ACIDS

25.1.1 Trivially named

Starting at the –COOH group, explore along the longest available uninterrupted †chain of carbon atoms, disregarding side-substituents. Does the chain with the –COOH group at one end correspond to any of the following?

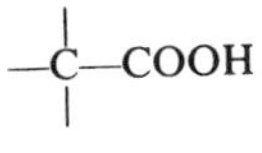

$$-\overset{|}{\underset{|}{C}}-COOH$$

Acetic acid

$$-\overset{|}{\underset{|}{C}}-\overset{|}{\underset{|}{C}}-COOH$$

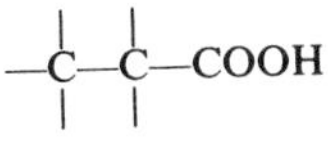

Propionic acid

† 'Uninterrupted' means without intervening heteroatoms or ring-systems.

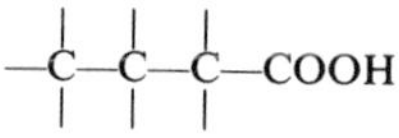

$$-\overset{|}{\underset{|}{C}}-\overset{|}{\underset{|}{C}}-\overset{|}{\underset{|}{C}}-COOH$$

Butyric acid

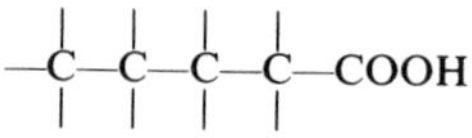

$$-\overset{|}{\underset{|}{C}}-\overset{|}{\underset{|}{C}}-\overset{|}{\underset{|}{C}}-\overset{|}{\underset{|}{C}}-COOH$$

Valeric acid

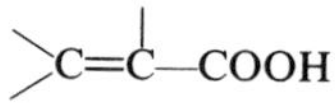

$$\overset{}{>}C=\overset{|}{C}-COOH$$

Acrylic acid

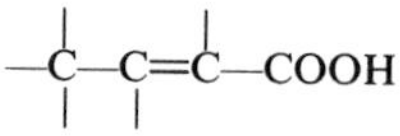

$$-\overset{|}{\underset{|}{C}}-\overset{|}{C}=\overset{|}{C}-COOH$$

Crotonic acid‡

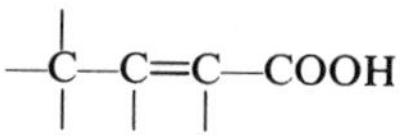

$$-\overset{|}{\underset{|}{C}}-\overset{}{C}=\overset{}{C}-COOH$$

Isocrotonic acid‡

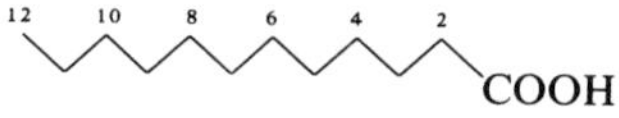

Lauric acid

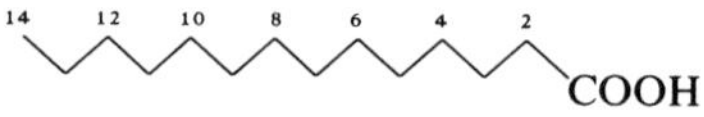

Myristic acid

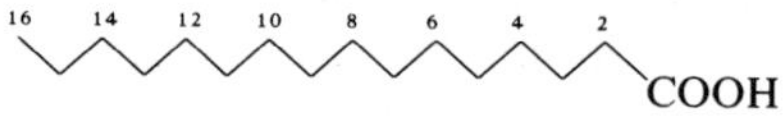

Palmitic acid

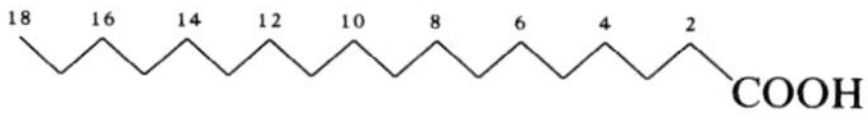

Stearic acid

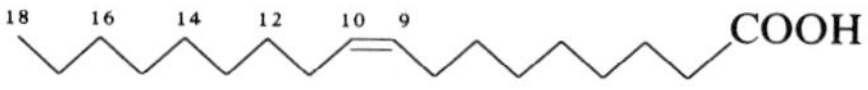

Oleic acid‡

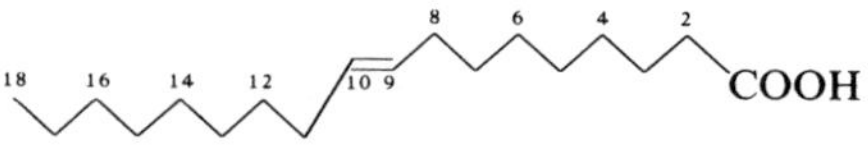

Elaidic acid‡

‡ In the special case of these five acids, the name covers the stereochemical situation and the provisions of section 2 do not apply.

The following is also retained so long as the side-chain retains its double bond but bears no alkyl group, and the ring is not hydrogenated:

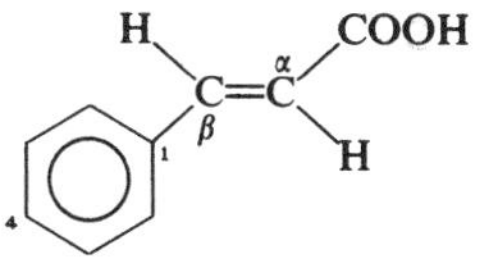

Cinnamic acid‡

If the answer to the question at the start of this section is 'YES',

then that name is the basis for naming substituted derivatives, all attached atoms or groups being cited in alphabetical order as radical prefixes (see Appendix C), each with appropriate locants, the first followed by a hyphen and the rest flanked by hyphens.

Example (1)

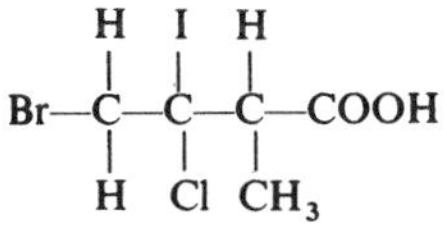

4-Bromo-3-chloro-3-iodo-2-methylbutyric acid
[prefix order: b,c,i,m]

Example (2)

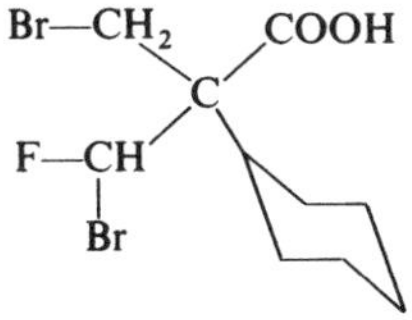

3-Bromo-2-bromomethyl-2-cyclohexyl-3-fluoropropionic acid

Here, there are two COOH-terminated chains of equal length. The one with more substituents is preferred. 'Bromo' precedes 'bromomethyl' in the name.

Groups occurring more than once directly attached to the main chain are collected in the name by means of prefixes di, tri, etc., or, if identically substituted themselves, by bis, tris, tetrakis, etc.

Example (3)

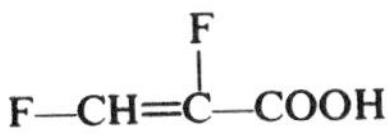

2,3-Difluoroacrylic acid

‡ The footnote on p.178 applies here.

Example (4)

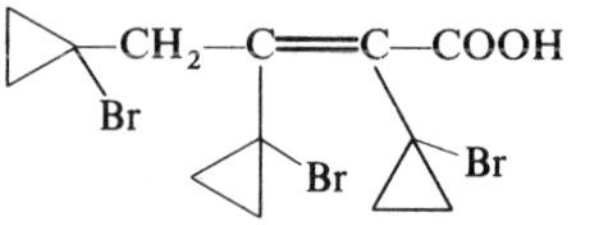

2,3,4-Tris(1-bromocyclopropyl)isocrotonic acid

Example (5)

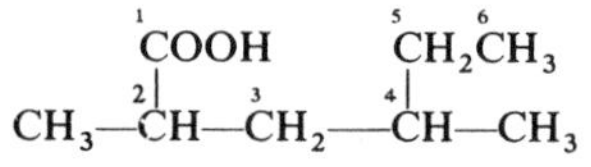

Here the presentation seems to put the –COOH group in mid-chain, but the PFS (see Appendix D-1.1) is identified by giving the –COOH carbon atom the locant 1, its adjacent carbon atom locant 2, and now locant 3 is assigned to the next carbon atom of the longer of the two branches, viz. the –CH$_2$– carbon atom. At the next branch-point the same applies and so the PFS is *not* one of the trivially named set.

25.1.2 Systematically named

If the answer to the question at the start of section 25.1.1 is 'NO',

then one of the following two methods are used, as appropriate:

(a) If the chain with –COOH at the end has no unsaturation, the name ends in '… anoic acid'. This is immediately preceded by a stem corresponding to the number of carbon atoms in the chain (including that of the –COOH group), constructed as follows:

   6: hex   7: hept   8: oct   9: non   10: dec   11: undec   12: dodec
   13: tridec   14: tetradec   15: pentadec   16: hexadec   17: heptadec
   18: octadec   19: nonadec   and 20: icos

Thus the PFS of example (5) of section 25.1.1 is hexanoic acid and its full name is 2,4-dimethylhexanoic acid.
   In the following case, however, the senior chain is decided not by criterion (B-8), yellow pages, but by (B-7).

Example (1)

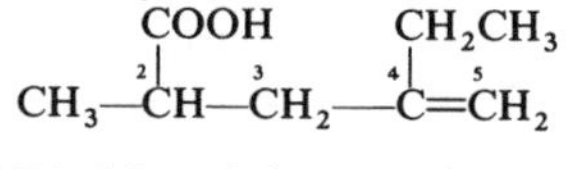

4-Ethyl-2-methylpent-4-enoic acid

Example (2)

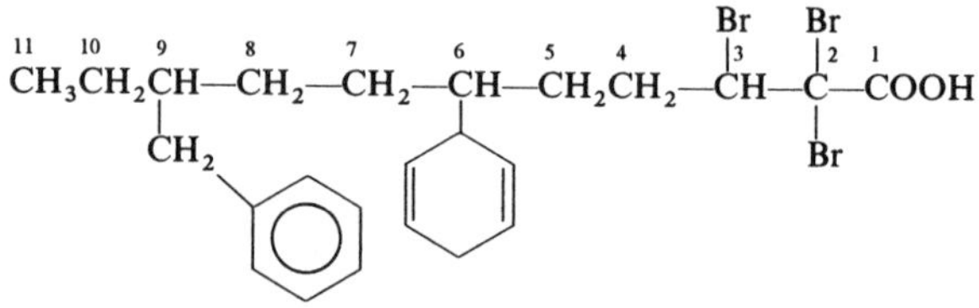

9-Benzyl-2,2,3-tribromo-6-cyclohexa-2,5-dienylundecanoic acid
[prefix-order: be,br, c]

(b) If the chain with the –COOH end has any unsaturation but is not one of the trivially named acids of section 25.1.1, the 'anoic' part of the name is modified to 'enoic' for one double bond, 'adienoic' for two, 'atrienoic', for three, etc.; 'ynoic' for one triple bond, 'diynoic' for two, etc.; 'enynoic' for one of each, etc. In each case, the locants for unsaturation are cited as described in section 5, the carbon atom of the –COOH group being counted as 1.

Example (3)

$$\underset{5}{CH_3}—\underset{}{\overset{}{C}}=\underset{4}{CH}—\underset{3}{CH_2}\underset{2}{CH_2}—\underset{1}{COOH}$$
$$\underset{6}{CH_2}$$

5-Methyl-6-phenylhex-4-enoic acid

Note. This is preferred to 5-benzylhex-4-enoic acid (using the alternative chain) because a chain carrying two substituents is preferred to a chain of otherwise equal seniority carrying only one. (see criterion (B-13), yellow pages).

Example (4)

$$CH_2CH_2CH_2CH_3$$
$$\underset{5}{I—CH_2}—\underset{4}{CH}=\underset{3}{CH}—\underset{2}{\overset{|}{C}}—\underset{1}{COOH}$$
$$CH_3—\overset{|}{CH}—CH_3$$

2-Butyl-5-iodo-2-isopropylpent-3-enoic acid

Example (5)

$$CH_3—\overset{}{CH}—CH_2—CH=CH—\underset{2}{\overset{}{CH}}$$
$$\overset{1}{COOH}$$
$$\underset{3}{CH}=\underset{4}{CH}—\underset{5}{CH}=\underset{6}{CH_2}$$

2-[4-(4-Pyridyl)pent-1-enyl]hexa-3,5-dienoic acid
[chain with two double bonds preferred to chain with only one]

25.2 PEROXY ACIDS

$$\overset{O}{\overset{||}{—C}}—O—OH$$

The name is formed exactly as if the group were –COOH, except that 'peroxy' is inserted immediately before the name of the unsubstituted acid.

Example (1)

$$Cl_3C{-}{-}CH(Cl){-}\overset{\overset{\displaystyle O}{\|}}{C}{-}O{-}OH$$

2,3,3,3-Tetrachloroperoxypropionic acid

Example (2)

$$Br{-}CH_2{-}CH{=}CH{-}CH{=}CH{-}\overset{\overset{\displaystyle O}{\|}}{C}{-}O{-}OH$$

6-Bromoperoxyhexa-2,4-dienoic acid

Exception

$$CH_3{-}\underset{\underset{\displaystyle O}{\|}}{C}{-}O{-}OH$$

is called peracetic acid.

25.3 MONOTHIOIC ACIDS

$$\overset{\overset{\displaystyle S}{\|}}{C}{-}OH \quad or \quad \overset{\overset{\displaystyle O}{\|}}{C}{-}SH$$

In the case of the free acids, these two forms are usually indistinguishable but either may be 'frozen' by ester-formation see section 25.8.2.

The monothioic acid equivalents of the first four acids listed under section 25.1.1 use the same names but preceded by 'thio', e.g.

$$CH_3CH_2CH_2C\overset{O}{\underset{S}{\diagup\!\!\!\diagdown}}{-}H$$

Thiobutyric acid

For others, the name of the corresponding hydrocarbon chain, obtained by replacing the $-COSH$ group by $-CH_3$ (see sections 5.1.1 and 5.1.2), is cited before the ending: 'thioic acid' for $-COSH$ and 'dithioic acid' for $-CSSH$. Numbering of the chain for unsaturation-locants and for siting any substituent groups (see Appendices C and D) begins with the acid-group carbon atom as 1.

Example (1)

$$CH_3{-}\overset{5}{C}{=}\overset{4}{C}H{-}\overset{3}{C}H_2\overset{2}{C}H_2\overset{1}{C}OSH$$
$$\underset{\overset{|}{\overset{6}{C}H_2}}{\bigcirc}$$

5-Methyl-6-phenylhex-4-enethioic acid
[cf. example (3) of section 25.1.2]

Example (2)

$$\underset{\text{2,3-Difluoropropenethioic acid}}{F-CH=\overset{\overset{\displaystyle F}{|}}{C}-\overset{\overset{\displaystyle S}{||}}{C}-OH}$$

'Thio' names usually need enclosing marks to avoid ambiguity as 'thio' can also mean –S–.

Example (3)

Chloro(piperidinothio)(pyrimidin-2-yl)(thioacetic acid)

25.4 DITHIOIC ACIDS

$CH_3-C(=S)-SH$ is named dithioacetic acid. This is exceptional; the normal method is covered under section 25.3 (q.v.).

25.5 SULPHONIC ACIDS

$$-C-\overset{\overset{\displaystyle O}{\uparrow}}{\underset{\underset{\displaystyle O}{\downarrow}}{S}}-OH$$

The name is based on the principal chain to which the $-SO_2OH$ group is directly attached, chosen on the same basis as in section 25.1.1 except that the PG does not have to occupy a terminal position on the chain (contrast $-COOH$). The chain is chosen and named according to section 5, and then, after a hyphen, the locant for the SO_3H group, then another hyphen and lastly 'sulphonic acid'.

If there is a choice, numbering is chosen which gives the lowest permissible locant to the SO_3H group.

Example

$$\overset{6}{CH_3}-\overset{5}{CH}=\overset{4}{CH}-\overset{3}{CH}-CH_2CH_2CH_2CH_2CH_3$$
$$\underset{1}{CH_3}-\overset{2}{CH}-SO_2OH$$

3-Pentylhex-4-ene-2-sulphonic acid

Of the possible PG-bearing chains, that with a double bond is preferred to that without. The PG site then gets the locant 2 rather than 5.

Sulphinic acids —C—S(=O)OH are named in exactly the same way as the sulphonic acids except that the stem 'phon' becomes 'phin'.

25.6 SALTS

$$\overset{\displaystyle O}{\underset{\displaystyle \|}{}} $$

$$-\overset{\|}{C}-OM, \quad -COSM, \quad -CSSM, \quad -SO_2OM$$

If the metal is monovalent, the name is formed by citing it and then, after a space, the anionic form of the appropriate acid, formed according to sections 25.1–25.5 (the ending 'ic acid' is changed to 'ate').

Ammonium NH_4 is treated (whether or not substituted) like a metal for the purposes of salt-naming.

Example (1)

$$F-CH=\overset{\displaystyle F}{\underset{\displaystyle |}{C}}-COOK$$

Potassium 2,3-difluoroacrylate

Example (2)

$$F-CH=\overset{}{C}-\overset{}{C}-SK$$
$$\underset{\displaystyle F}{|} \quad \underset{\displaystyle O}{\|}$$

Potassium S-(2,3-difluoropropenethioate)

Note the use of S (or O) locants to make the distinction normally unnecessary for the free acids, cf. section 24.4. For salts of dithio acids, these locants are omitted.

If the metal has a valency of more than 1, its appropriate ratio to the anion is indicated by means of appropriate multiplicative prefixes: di, tri, etc., or, if identically substituted, bis, tris.

Example (3)

$$Zn(CH_3COO)_2$$

Zinc di(acetate)

The parentheses here are to avoid possible confusion with salts of acetoacetic acid. which has loosely been called 'diacetic acid'.

Example (4)

$$Al^{3+} \left[O \underset{}{\bigcirc} N-CH_2CH_2COO^- \right]_3$$

Aluminium tris(3-morpholinopropionate)

25.7 CARBOXYLIC ANHYDRIDES

$$R^1 \overset{\overset{\displaystyle O}{\|}}{C} - O - \overset{\overset{\displaystyle O}{\|}}{C} - R^2$$

The central –O– atom is removed and –OH added to the two fragments so formed, to give R^1—COOH and R^2—COOH, respectively.

These acids are then identified and named as under section 25.1.1. They are then cited in alphabetical order, but without the word 'acid'—separated by a space. Finally, after another space, comes the word 'anhydride'.

If R^1 and R^2 are the same, the acid name (minus the word 'acid') is cited, then a space and then 'anhydride' [no 'di' prefix].

Example (1)

$$CH_3 - [CH_2]_4 - \overset{\overset{\displaystyle O}{\|}}{C} - O - \overset{\overset{\displaystyle O}{\|}}{C} - CH_2CH_3$$

Hexanoic propionic anhydride

Example (2)

$$CH_2{=}CH - \underset{\underset{\displaystyle O}{\|}}{C} - O - \underset{\underset{\displaystyle O}{\|}}{C} - CH{=}CH_2$$

Acrylic anhydride

25.8 ESTERS

$$-COOR, \ -C(=S)-OR, \ C(=O)-SR, \ -CSSR, \ -SO_2-OR$$

25.8.1 Esters of carboxylic acids

$$-COOR$$

The group R is expressed as a radical (see Appendix C) and then, after a space comes the anionic form of the acid exactly as for salts (see section 25.6). The only difference in name construction, apart from that between metal and radical, is that any O or S locants precede the radical rather than the anion.

Example

4-Bromophenethyl 2-bromopropionate

25.8.2 Esters of carbothioic acids

$$—C(=S)—OR \quad or \quad —C(=O)—SR$$

The first form are named by citing in order: '*O*-', then the radical prefix for the non-acid moiety; then a space and then the anion from the appropriate acid ('-ic acid' becomes '-ate').

In the case of the second form, *S*- replaces *O*-.

Example (1)

$$CH_3—CH_2CH_2—\underset{\underset{S}{\|}}{C}—O—CH_2Cl$$

O-Chloromethyl thiobutyrate

Example (2)

$$CH_3—CH_2—CH=CH—[CH_2]_6—\overset{\overset{S}{\|}}{C}—OCH_2—O—CH_3$$

O-Methoxymethyl undec-8-enethioate

Example (3)

$$\text{(naphthyl)}—\underset{\underset{O}{\|}}{\overset{S}{C}}—CH=CHF$$

S-(2-Naphthyl) 3-fluoropropenethioate

Example (4)

$$CH_3—[CH_2]_4—\overset{\overset{O}{\|}}{C}—S—CH_2CH_3$$

S-Ethyl hexanethioate

Example (5)

$$Cl_2C=CH—\underset{\underset{S}{\|}}{C}—S—CH_2CH_3$$

Ethyl 3,3-dichloro(dithioacrylate)

25.8.3 Esters of sulphonic acids

$$R—O—SO_2—R'$$

The name is made up of the following parts cited in the order given: (a) the group R in radical form (see Appendix C), (b) a space, (c) the group R' named as if the $-OSO_2R$ were replaced by H, (d) the locant of attachment of the $-SO_2OR$ group, (the lower if there is a choice), and (e) 'sulphonate'.

Example (1)

$$CH_3CH_2CH=CH-CH-[CH_2]_7-CH_3$$

$$\quad\quad\quad\quad\quad O-SO_2-CH-CH_2-CH=CH-CH_3$$

$$\quad\quad\quad\quad\quad\quad\quad\quad CH_2CH_3$$

1-Octylpent-2-enyl hept-5-ene-3-sulphonate
[for numbering see Appendix D-2]

25.8.4   Esters of inorganic acids†

$$R-O-[Q](=O)_x(OH)_y$$

Where Q is the central element, usually a non-metal (e.g. N,S,P), and x and y are integers (possibly 0).

See under alcohols (section 25.2.3).

25.9   LACTONES

See section 24.9.

25.10   ACYL HALIDES

$$\overset{\displaystyle O}{\underset{\displaystyle R-C-X}{\|}}$$

Where X is F, Cl, Br or I.

In the case of the acids (5)–(13) of section 25.1.1 the names are formed by changing the ending 'ic acid' to 'oyl halide'. In all other cases, including systematically named acids, the corresponding change is from 'ic acid' to 'yl halide'.

Example (1)

$$CH_3C(=O)-Cl$$

Acetyl chloride (from acetic acid)

Example (2)

$$H_2C=CH-C(=O)-Br$$

Acryloyl bromide (from acrylic acid)

---

† Carbonic acid is included as an inorganic acid (dibasic) $O=C(OH)_2$, as is also the hypothetical orthocarbonic acid $(C(OH)_4$. In each case their esters are treated under –OH compounds, see sections 21.7 and 25.20.3.

Example (3)

$$CH_3[CH_2]_4CH=CH-C(=O)-I$$

Oct-2-enoyl iodide (from oct-2-enoic acid)

For naming substituted derivatives, the numbering is that of the parent acid (see Appendix C for radical prefix names).

25.11 CARBOTHIOIC HALIDES

$$-C(=S)-X$$

If the acid derives from one of the trivially named structures of section 25.1.1, then the name is formed as in section 25.10 but is preceded by 'thio'. For acids otherwise derived, the ending '-oic acid' of the corresponding section 25.1.2 name is changed to 'ethioyl chloride', '-ethioyl bromide', etc. as appropriate.

Example (1)

$$CH_3C(=S)-I$$

Thioacetyl iodide

Example (2)

$$CH_3[CH_2]_5-C(=S)-F$$

Heptanethioyl fluoride

25.12 HALIDES OF SULPHONIC ACIDS

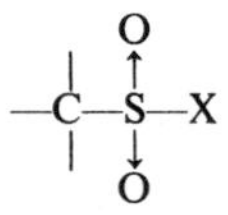

The name is formed by changing the ending of the name of the corresponding acid from 'ic acid' to 'yl' and adding, after a space, the anion-name from the appropriate halogen.

Example

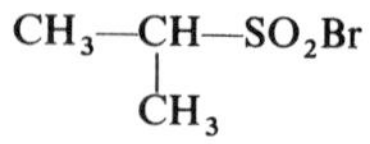

$$CH_3-CH-SO_2Br$$
$$\quad\quad\ |$$
$$\quad\quad CH_3$$

Propane-2-sulphonyl bromide

25.13 UREAS

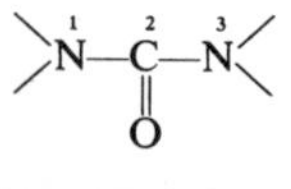

These are named as described in section 24.11, except that all the radicals directly attached to the nitrogen atoms are acyclic chains.

The group

$$>N-\underset{\underset{S}{\|}}{C}-N<$$

is called 'thiourea' It is numbered 1–3 as for urea but it is enclosed inside parentheses whenever there is any possible ambiguity due to the alternative use of 'thio' to mean a –S– group in a chain.

Example

$$CH_3-\underset{\underset{CH_2CH_3}{|}}{N}-\underset{\underset{S}{\|}}{C}-NH-CH_2CH_3$$

1,3-Diethyl-1-methyl(thiourea)

25.14 AMIDES

$$-C(=O)-NH_2; \quad -SO_2NH_2$$

The name is formed by changing that for the corresponding acid of sections 25.1–25.5. Endings 'ic acid' or 'oic acid' become 'amide'.

If there are substituent groups attached to the nitrogen atom, they are cited as radical prefixes, preceded by the N-locant and they take their place in the alphabetic sequence.

Example (1)

$$Br_2C=\underset{\underset{NO}{|}}{C}-\underset{\underset{O}{\|}}{C}-N\overset{CH_3}{\underset{Br}{}}$$

N,3,3-Tribromo-*N*-methyl-2-nitrosoacrylamide

Example (2)

$$\underset{CH_3}{\overset{CH_3}{>}}CH-CH_2CH_2\underset{\underset{SO_2-N(CH_3)_2}{|}}{CH}-CH_3$$

N,*N*,5-Trimethylhexane-2-sulphonamide

Example (3)

$$\overset{5}{C}H_3-\overset{4}{C}H=\overset{3}{\underset{\underset{\overset{1}{C}H_3}{|}}{C}}-\overset{2}{\underset{\underset{CH_3}{|}}{C}H}-SO_2NH-CH_3$$

N,3-Dimethylpent-3-ene-2-sulphonamide

Structures having a benzene ring attached to the amide nitrogen atom are named by replacing 'amide' in the appropriate –$CONH_2$ name by 'anilide'. Numbering is unchanged for the chain but the ring-positions have 'primed' numerals, 1′ being the point of attachment to the nitrogen atom.

Example (4)

$$BrCH_2CH_2C{-}NH{-}\text{(ring)}{-}Br$$

2′,3,4′-Tribromopropionanilide

25.15   AMIDINES

$$-C(=NH)-NH_2$$

These are regarded as being derived from the acids of section 25.1, the above group having replaced the –COOH group. Accordingly, their names are derived by citing that of the corresponding acid and replacing 'acid' or 'oic acid', as the case may be, by 'amidine'.

Example (1)

$$CH_3-C(=NH)-NH_2$$

Acetamidine

Example (2)

$$CH_3-[CH_2]_4-C(=NH)-NH_2$$

Hexanamidine

Substituted derivatives

For these, groups attached to the –NH are given the locant N^1 whilst those attached to the =NH are given the locant N^2.

25.16   NITRILES

$$R-CN$$

If –CN replaces –COOH in any of the trivially named acids of section 25.1.1, the name is formed by replacing 'ic acid' by 'onitrile', e.g. valeric acid—valeronitrile for $CH_3CH_2CH_2CH_2CN$.

For others, the senior chain with a terminal –CN group is chosen and named as if it were a –CH_3 group. 'Nitrile' is then added.

In naming substituted derivatives, the carbon atom of the CN group carries the number 1.

Example

$$\overset{6}{Br-CH_2}\overset{5}{CH_2}\overset{4}{CH}=\overset{3}{CH}-\overset{2}{CH_2}\overset{1}{CN}$$

6-Bromohex-3-enenitrile

25.17   ALDEHYDES

$$R-CHO$$

If derived from any of the trivially named acids of section 25.1.1, the name is formed by changing 'ic acid' to 'aldehyde'. If not, the senior chain is chosen as if the terminal –CHO group were a $-CH_3$ group, named according to section 5, and 'al' added after elision of the terminal 'e'.

Substituent locants are assigned on the basis that the –CHO group is numbered 1.

Example (1)

$$CH_3 - CHI - CHO$$

2-Iodopropionaldehyde

Example (2)

$$CH_3-CH=CH-CH_2-CH=\overset{\overset{\displaystyle CH_3CH_2}{|}}{C}-CHO$$

2-Ethylhepta-2,5-dienal

25.18   KETONES

$$R^1-\underset{\underset{\displaystyle O}{\|}}{C}-R^2$$

Where R^1 and R^2 are alkyl groups.

The structure $CH_3-C(=O)CH_3$ is named 'acetone'.

For others, the senior chain bearing the non-terminal $-C(=O)-$ group is chosen and named as if this group were a $-CH_2-$ group instead (according to section 5) but dropping the terminal 'e'. Next comes the locant for the $-C(=O)-$ group, flanked by hyphens and chosen to be as low as permissible, and lastly 'one'.

Example (1)

$$\overset{1}{CH_3}\overset{2}{CH_2}\overset{\overset{\displaystyle O}{\|}}{C}-\overset{4}{CH_2}\overset{5}{CH_2}\overset{6}{CH}=\overset{7}{CH_2}$$

Hept-6-en-3-one

Example (2)

$$CH_3O-\overset{1}{C}H_2\overset{2}{C}H_2\overset{\overset{\displaystyle O}{\|}}{\overset{3}{C}}-\overset{4}{C}H_2\overset{5}{C}H_3$$

1-Methoxypentan-3-one

The $-C(=O)-$ group is numbered 3 from either end. 1 is lower than 5 for the methoxy group.

25.19 THIOKETONES

$$R^1-\overset{\overset{\displaystyle }{|}}{\underset{\underset{\displaystyle S}{\|}}{C}}-R^2$$

The naming procedure is as for ketones, except that the trivial names are dropped, the final 'e' of the chain-name is restored, and the ending is 'thione'.

Example

$$CH_3-\overset{\overset{\displaystyle S}{\|}}{C}-CH_2CH_2CH_3$$

Pentane-2-thione

25.20 ALCOHOLS, THEIR METAL DERIVATIVES AND ESTERS WITH MINERAL ACIDS

$$>C=N-OH$$

25.20.1 Alcohols

(Other than 'radico-alcohols' of section 21.6, which are preferred if appropriate.)

The name is formed by citing the name of the senior chain (see section B, yellow pages) bearing the –OH group (which may or may not be attached to a terminal carbon atom), deleting the final 'e' and adding the lower locant for the –OH group, flanked by hyphens, and then 'ol'. Methanol and ethanol have no locants when unsubstituted and methanol needs none even when substituted. Attached groups are cited as radical prefixes (see Appendix C) in alphabetical order.

Numbering

The lower locant is assigned to the position of attachment of the –OH group on the senior chain. Any unsaturation locants are assigned after that by sequential numbering.

Example (1)

$$CH_3CH-CH_2-CH_3$$
$$CH_3-[CH_2]_4-\overset{\displaystyle |}{C}HOH$$

3-Methylnonan-4-ol
(Not 2-ethyloctan-3-ol, nor 7-methylnonan-6-ol)

Example (2)

$$F_3C-CH_2-CH-CH_2OH$$
$$Br_2CH-CH-Cl$$

4,4-Dibromo-3-chloro-2-(2,2,2-trifluoroethyl)butan-1-ol
[prefix-order; b,c,t]

Here, the two possible chains are of equal length, carry the same degree of unsaturation and have the same number of substituents. Seniority is decided on the lowest locant-set: 2,3,4,4 rather than 2,4,4,4.

Example (3)

3-Cyclopent-2-enyl-4-methylhex-5-en-3-ol
[not ... -1-en-4-ol]

25.20.2 Metal derivatives of chain alcohols

In the case of monovalent metals the name consists of that of the metal, a space, the name of the alkyl radical, another space and then "oxide".

Example (1)

$$C_6H_5-CH_2-O-Na$$

Sodium benzyl oxide

In the case of polyvalent metals the name of the metal is followed by a space, then 'bis' or 'tris' etc., as appropriate, open parenthesis, the name of the radical, a space, then "oxide" and closed parenthesis.

For restrictions on the use of trivial names for branched-chain radicals and selection of the main branch, see Appendix C-1.1.

Example (2)

$$Zn \left[O-CH(Bu)-CH=CH_2 \right]_2$$
$$Zn$$

Zinc bis(1-butylallyl oxide)

As exceptions the contracted 'alkoxy' radicals listed in Appendix C-2.2 are used. The name in these cases consists of that of the metal, a space, and then the name of the appropriate alkoxide. For polyvalent metals the alkoxide name is preceded by di, tri, etc. for unsubstituted radicals and bis, tris, etc. for substituted radicals. If such substituents include C-chain attachment, the restrictions of Appendix C-1.1, (q.v.) apply. Names for branched radicals, e.g. *tert*-butoxy, may not be used when bearing **any** substituents. Their use requires a hyphen before as well as after "*sec*" or "*tert*", but not 'iso'.

Example (3)

$$Ti \left[- O - CH(CH_3)_2 \right]_4$$

Titanium tetraisopropoxide

Example (4)

$$Ba \left[- O - C(CH_3)_3 \right]_2$$

Barium di-*tert*-butoxide

Example (5)

$$CH_3 - CH - OK$$
$$\quad\quad\quad | $$
$$\quad\quad\quad CH_2Cl$$

Potassium 2-chloro-1-methylethoxide
[Substituted isopropoxide. Vertical chain preferred to horizontal by the principle of criterion B-(13),
Yellow pages.]

25.20.3 Esters of chain alcohols with mineral acids

The name of the appropriate radical is followed by a space and then the name of the anion. Carbonates and orthocarbonates are included in this set. In the case of polybasic acids, the prefixes di, tri,... and bis, tris, ... are used, as prescribed in section 25.20.2, q.v.

Acid esters are named by prefixing the name of the anion with "hydrogen" or "dihydrogen" as appropriate.

Example (1)

$$\begin{array}{c} O \\ \| \\ CH_3CH_2 - O - S - O - CH_2CH_3 \end{array}$$

Diethyl sulfite

Example (2)

$$\begin{array}{c} CH_3\,CH_2 \\ | \\ CH_3CH_2CH_2CH - O - COOH \end{array}$$

1-Ethylbutyl hydrogencarbonate
[NB No space after 'hydrogen' in the case of 'inorganic' acids.]

In the case of mixed esters, the radicals are cited in alphabetical order, separated by a space, another space, then the anion-name (preceded by 'hydrogen' if appropriate).

Example (3)

$$\begin{array}{c} OH \quad\quad\quad CH_2 \\ | \quad\quad\quad\quad \| \\ CH_3CH_2CH_2 - CH - O - P - O - C - CH_2CH_2CH_2CH_3 \\ | \\ CH_2CH_3 \end{array}$$

1-Butylvinyl 1-ethylbutyl hydrogenphosphite

25.21 THIOLS

All are named as under section 25.20.1 except that the final 'e' of the senior chain-name is retained and the ending 'ol' is replaced by '-thiol'.

Numbering

This is as for the alcohols of section 25.20.1 and the same considerations apply regarding locants for the methane and ethane cases as for methanol and ethanol.

Example (1)

$$CH_3CH_2CH_2—SH$$

Propane-1-thiol

Example (2)

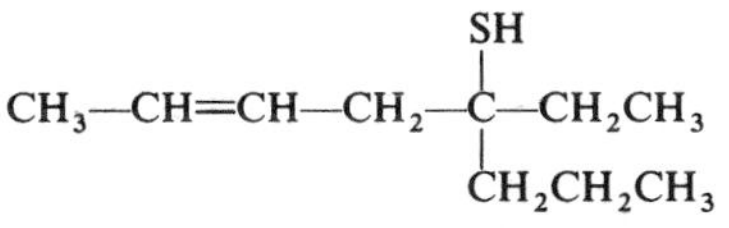

$$CH_3—CH=CH—CH_2—\overset{\displaystyle SH}{\underset{\displaystyle CH_2CH_2CH_3}{C}}—CH_2CH_3$$

4-Ethyloct-6-ene-4-thiol

Example (3)

Example (3) of section 26.0 (q.v.) is named

[2,3-Bis(mercaptomethoxy)-2-(2-mercaptopropoxymethyl)propane-1-thiol]

The alternative PFS* also a propane-1-thiol, is eliminated by criterion (B-15), yellow pages ('mercaptomethoxy' preferred alphabetically to 'mercaptomethoxymethyl').

*Defined in section 1.5. See also Appendix D-1.1.

26 Poly-PG acyclics

26.0 PREAMBLE

Appendix D should be studied and the structure examined with a view to collecting as many PGs in the name-ending as possible.

If this is the case for two or more groups, then sections 26.1.1, 26.1.2, 26.1.5, 26.2 and 26.4, as appropriate, should be searched for appropriate trivially named parent-structures which fit the case when stripped of substituent groups.

If none is applicable, systematic methods are used. Consult the Table of contents and turn to the relevant part of this section.

The applicability of section 26.4 should be explored before proceeding. Where the same PFS*, disregarding any substituent groups, occurs more than once in a structure, separated by an atomic or molecular group capable of being named as a divalent or polyvalent radical, section 26.4 (q.v.) is appropriate, provided that the seniority rules (in particular criterion (B-1), yellow pages) permit.

Thus, in example (1) of this sub-section the PFS* is the portion below the dotted line, because the name butane-1,2,4-tricarbonitrile collects three PGs (defined in section 1.5), whereas, although an assembly name ending in '. . . oxydimalononitrile' might seem to collect four PGs, 'malononitrile' is the PFS and it collects only two.

Example (1)

$$
\begin{array}{c}
\text{NC—CH—CN} \\
| \\
\text{O} \\
| \\
\text{NC—C—CN} \\
\cdots\cdots\cdots\cdots\cdots\cdots | \cdots\cdots\cdots\cdots\cdots\cdots \\
\underset{4}{\text{NC—CH}_2}\text{—}\underset{3}{\text{CH}}\text{—}\underset{2}{\text{CH}}\text{—}\underset{1}{\text{CH}_2}\text{—CN} \\
| \\
\text{CN}
\end{array}
$$

*Defined in section 1.5. See also Appendix D-1.1.

Apart from specific instructions to the contrary where they occur in section 26, the procedures for name construction are based on those for the corresponding single-PG molecules of section 25, except that the prefixes di,tri, etc., as appropriate, are placed before the name of the PG.

Example (2)

$$CH_2-CH_2-OCH_2OH$$

$$\overset{11}{CH_3}-\overset{10}{CHOH}-\overset{9}{CH_2}\overset{8}{CH}=\overset{7}{CH}-\overset{6}{CHOH}-\overset{5}{CH_2}-\overset{4}{CHOH}-\overset{3}{CH_2}-\overset{2}{\underset{|}{CH}}-\overset{1}{CH_2OH}$$

2-[2-(Hydroxymethoxy)ethyl]undec-7-ene-1,4,6,10-tetrol
(The 'a' of 'tetra' is elided before 'ol'.)

It is impossible to make a name using all five –OH groups in the ending of example (2).

Example (3)

$$O-CH_2-SH$$

$$HS-CH_2\overset{|}{C}-CH_2-O-CH_2-SH$$

$$\underset{|}{CH_2}-O-CH_2-\underset{|}{CH}-CH_3$$

$$SH$$

Here, no collection of the 'thiols' is possible and the appropriate name is given under Ex.(3) of section 25.21.

26.1 POLYBASIC CARBOXYLIC ACIDS

26.1.1 Trivially named saturated dicarboxylic acids

The following are retained as PFSs (see Appendix D-1.1), numbered as shown. Their use is preferred to more systematic names and they are senior, in the event of a choice of PFS, to other dicarboxylic acids. For designation of individual stereoisomers, see section 22.2.5.

$$NH_2-\overset{2}{CH}-\overset{1}{COOH}$$
$$\overset{3}{CH_2}-\overset{4}{COOH}$$

(1) Aspartic acid

$$NH_2-\overset{2}{CH}-\overset{1}{COOH}$$
$$\overset{3}{CH_2}\overset{4}{CH_2}\overset{5}{COOH}$$

(2) Glutamic acid

When a chain carries two COOH groups the following trivial names are used as the basis of the name for substituted derivatives:

COOH
|
COOH

(3) Oxalic acid†

COOH
/
CH₂
\
COOH

(4) Malonic acid

† Derivatives of oxalic acid will take the form of salts and esters.

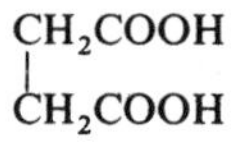

CH$_2$COOH
|
CH$_2$COOH

(5) Succinic acid

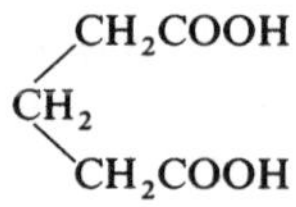

CH$_2$COOH
/
CH$_2$
\
CH$_2$COOH

(6) Glutaric acid

CH$_2$CH$_2$COOH
|
CH$_2$CH$_2$COOH

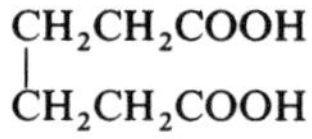

(7) Adipic acid

Numbering proceeds sequentially
from one COOH to the other.

The following is also retained:

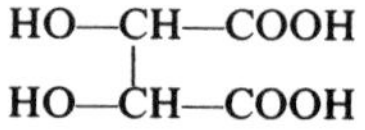

HO—CH—COOH
|
HO—CH—COOH

(8) Tartaric acid

and the name is used as a basis for naming *O*-acyl substituents, but other *O*-substituted derivatives are named on the basis of succinic acid.

Example (9)

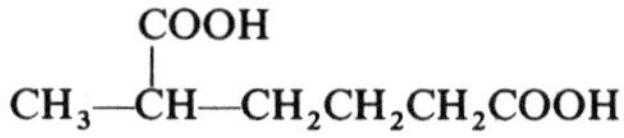

COOH
|
CH$_3$—CH—CH$_2$CH$_2$CH$_2$COOH

2-Methyladipic acid

Example (10)

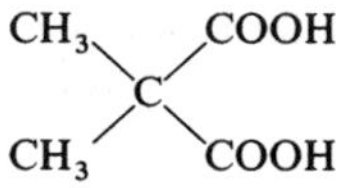

CH$_3$ COOH
\   /
C
/   \
CH$_3$ COOH

Dimethylmalonic acid
[no locants needed]

Substituted derivatives assign the lowest possible locant-set to the radical prefixes, cited in alphabetical order before the name of the acid.

Example (11)

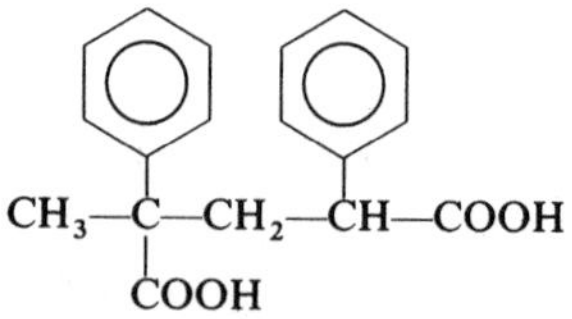

CH$_3$—C—CH$_2$—CH—COOH
|
COOH

2-Methyl-2,4-diphenylglutaric acid

26.1.2 Trivially named unsaturated dicarboxylic acids

The following are used as a basis for naming derivatives:

$$
\begin{array}{cc}
\text{H—C—COOH} & \text{HOOC—C—H} \\
\text{‖} & \text{‖} \\
\text{H—C—COOH} & \text{H—C—COOH}
\end{array}
$$

(1) Maleic acid   (2) Fumaric acid

These names are sterically specific and need no other descriptors.

If either of these is substituted by a carbon chain, however short, the systematic name is preferred, viz. (Z)-[or (E)-]but-2-enedioic acid, respectively [see section 2.2 for the significance of Z and E].

26.1.3 Other dicarboxylic acids

Where both COOH groups are on the same chain of uninterrupted carbon atoms, they are regarded as occupying terminal positions. The name is formed by citing that of the chain of the appropriate carbon-number, including both COOHs (see section 5) and adding 'dioic acid'.

Example (1)

$$
\overset{1}{\text{C}}\text{OOH} \qquad \overset{7}{\text{C}}\text{OOH}
$$
$$
\underset{2\ \ \ \ 3}{\text{H}\overset{}{\text{C}}{=}\text{CH}}\text{—}\underset{4\ \ \ \ 5}{\text{CH}{=}\text{CH}}\text{—}\underset{6}{\text{CH}}\text{—CH}_3
$$

6-Methylhepta-2,4-dienedioic acid
[2,4-preferred to 3,5- for the unsaturation locants]

Chains bearing two $C({=}O)SH$ [$C({=}S)OH$] groups are named by treating both these sites as chain-ends and citing the name of the chain formed when unsubstituted (see sections 5.1.1 and 5.1.2), then adding 'bis(thioic acid)'. If both are —$C({=}S)$—SH groups, the same procedure is followed but the ending is instead 'bis(dithioic acid)'.

Example (2)

$$
\text{H}\cdot\underset{\text{S}}{\overset{\text{O}}{\diagdown}}\text{C—[CH}_2]_4\text{—C}\overset{\text{O}}{\underset{\text{S}}{\diagup}}\cdot\text{H}
$$

Hexanebis(thioic acid)

Example (3)

$$
\text{HS—CS—[CH}_2]_4\text{—CS—SH}
$$

Hexanebis(dithioic acid)

26.1.4 Acids with three or more –COOH groups

The name of the chain carrying the maximum number of COOH groups is followed by 'tri' or 'tetra', etc., and then 'carboxylic acid'. Any COOH groups on other chains are expressed as carboxy-substituents (see the entry under 'Collected', section 1.5).

Example

$$
\begin{array}{c}
\overset{\text{COOH}}{|} \quad \overset{\text{COOH}}{|} \quad \overset{\text{COOH}}{|} \quad \overset{\text{COOH}}{|} \\
\overset{1}{\text{CH}_3}-\overset{2}{\text{CH}}-\overset{3}{\text{CH}}_2-\overset{4}{\text{C}}-\overset{5}{\text{CH}}_2-\overset{6}{\text{CH}}-\overset{7}{\text{CH}}=\overset{8}{\text{C}}-\overset{9}{\text{CH}}_2\overset{10}{\text{CH}}_3 \\
\quad\quad\quad\quad\quad | \\
\quad\quad\quad\quad\text{CH}_2-\text{COOH}
\end{array}
$$

4-(Carboxymethyl)dec-7-ene-2,4,6,8-tetracarboxylic acid

26.1.5 Carboxylic acid derivatives

For salts and esters the name begins with the metal (or **cation**) or the appropriate radical as the case may be and then, after a space, follows the appropriate acid name converted to its anion form by replacing '-ic acid' with '-ate'.

For acid salts and acid esters see section 22.2.0.

The transformations in respect of acyl halides, amides, amidines, nitriles and aldehydes depend on whether the senior chain bears two PGs or more than two (contrast name and numbering of ex(1) of section 26.1.3 with those of the example to section 26.1.4). In column 2 of the following table the stem of the trivially named parent acid precedes the ending shown–without a break. In column 3 the ending shown is preceded by the name from section 5 of the senior hydrocarbon chain, except that locants for unsaturation follow those chosen as the lower set for the PGs.

PG	Trivial ending	Systematic ending
—COOH	-ic acid	dioic acid
		tri(tetra)carboxylic acid
—COX	-yl dihalide	dioyl dihalide
		tri(tetra)carbonyl (tri(tetra))halide
—CON<	-amide	diamide
		tri(tetra)carboxamide
N— ‖ —C—N<	-amidine	diamidine
		tri(tetra)carboxamidine
—CN	-onitrile	dinitrile
		tri(tetra)carbonitrile
—CHO	-aldehyde	dial
		tri(tetra)carbaldehyde

Examples

$$
\begin{array}{c}
\text{CH}_2-\text{CHO} \\
| \\
\text{CH}_2-\text{CHO}
\end{array}
$$

(1) Succinaldehyde

$$
\begin{array}{c}
\quad\;\text{CH}_2\text{CONH}_2 \\
\diagup \\
\text{CH}_2 \\
\diagdown \\
\quad\;\text{CH}_2\text{CONH}_2
\end{array}
$$

(2) Glutaramide

$$
\begin{array}{c}
\quad\;\overset{\text{O}}{\overset{\|}{\text{C}}}-\text{Br} \\
\diagup \\
\text{CH}_2 \\
\diagdown \\
\quad\;\underset{\|}{\text{C}}-\text{Br} \\
\quad\;\;\text{O}
\end{array}
$$

(3) Malonyl dibromide

$$
\begin{array}{c}
\text{HC}-\text{COO}^- \\
\|\quad\quad\quad\quad \text{Ca}^{2+} \\
\text{HC}-\text{COO}^-
\end{array}
$$

(4) Calcium maleate

H$_2$N—CH—COONa
|
CH$_2$CH$_2$COONa

NC—[CH$_2$]$_7$—CN

(5) Disodium glutamate

(6) Nonanedinitrile

CH$_3$—O—C—CH$_2$—CH—[CH$_2$]$_4$—C—S—CH$_2$CH$_3$
‖ | ‖
S   Cl   O

(7) *S*-Ethyl *O*-methyl 3-chloro-octanebis(thioate)

For situations where only one COOH group of such an acid is so transformed, see section 22.

26.1.6 Sulphonic acids and their derivatives

$$-SO_2OH, -SO_2X, -SO_2NH_2$$

Names are formed as for the corresponding PGs in sections 25.5, 25.12, and 25.14 (q.v.). These groups may occupy any chain position and the senior chain to which they are attached is first identified and named according to sections 5.1.1 or 5.1.2. Lowest locants are then given to the PGs as a set.

Example (1)

SO$_2$OH
|
CH$_3$—CH—CH—CH—CH$_2$CH$_3$
| |
SO$_2$OH   SO$_2$OH

Hexane-2,3,4-trisulphonic acid (Not 3,4,5-)

Example (2)

CH$_3$—CH=CH—CH—CH$_2$CH$_2$CH$_2$SO$_2$NH$_2$
|
SO$_2$NH$_2$

Hept-5-ene-1,4-disulphonamide

When the nitrogen atoms bear substituents, the radical name for each attached group is cited before the name of the parent structure (see Appendices C and D). The nitrogen atoms are distinguished by superscript numerals for the chain position added to the N-locants. *N*-phenyl substituents are named with the ending 'sulphonanilide' and ring-locants carry primes.

Example (3)

SO$_2$NHCH$_3$ SO$_2$NH$_2$
| |
Cl—CH$_2$CH$_2$CH—CH—CH—[CH$_2$]$_3$—CH—CH$_3$
| |
CH$_3$ SO$_2$NH$_2$

10-Chloro-N^8,7-dimethyldecane-2,6,8-trisulphonamide

26.1.7 Other acids

Sulphinic acids, —S(O)—OH, are named as under section 26.1.6 except that the ending 'inic' replaces 'onic', etc.

Selenium analogues of these acids also follow the procedures of section 26.1.6 but 'sulphonic' and 'sulphinic' are replaced by 'selenonic' and 'seleninic'.

26.2 KETONES

The compound

$$CH_3—\overset{\overset{\displaystyle O}{\|}}{C}—\overset{\overset{\displaystyle O}{\|}}{C}—CH_3$$

is called biacetyl but, in general, where a compound carries two or more —C(=O)— groups, the name for the chain carrying the maximum number of these groups is cited according to section 5 and then the keto-groups collected together with their locants, the prefixes di, tri, etc., as appropriate, and the ending 'one'.

The locant-set for the =O groups is as low as possible. If it is the same starting from either end of the senior chain, any unsaturation takes the lower locant set.

Example (1)

$$CH_3—\overset{\overset{\displaystyle O}{\|}}{C}—CH=CH—\overset{\overset{\displaystyle O}{\|}}{C}—CH—CH_2—\overset{\overset{\displaystyle O}{\|}}{C}—CH_3$$
$$CH_3—\overset{\overset{\displaystyle O}{\|}}{C}—CH_2—[CH_2]_3—\overset{\displaystyle |}{C}H_2$$

6-Acetonyltridec-3-ene-2,5,12-trione

Example (2)

$$CH_3—\overset{\overset{\displaystyle O}{\|}}{C}—CH=CH—\overset{\overset{\displaystyle O}{\|}}{C}—CH_2—CH_2—\overset{\overset{\displaystyle O}{\|}}{C}—CH_3$$

Non-3-ene-2,5,8-trione
[not -6-ene]

If one or more >C=O positions of the chain is joined directly to a ring, the locant 1 is used in the suffix-set. See also ex.(3) of section 27.4.2.

26.3 POLYOLS

The following trivial names are retained, but are not used for naming substituted derivatives [see Example (6)].

$$CH_2OH$$
$$|$$
$$CH_2OH$$

(1) Ethylene glycol

$$CH_2OH$$
$$|$$
$$CH-OH$$
$$|$$
$$CH_3$$

(2) Propylene glycol

$$CH_2OH$$
$$|$$
$$CH-OH$$
$$|$$
$$CH_2OH$$

(3) Glycerol

(4) Pentaerythritol

Other polyols are named by citing the section 5 name for the chain bearing the greatest number of –OH groups, followed by 'diol' or 'triol' etc. Any others are cited as hydroxy-prefixes.

Example (5)

2,2-Bis(ethoxymethyl)propane-1,3-diol

Example (6)

2-(Hydroxymethyl)butane-1,2,4-triol

If there is a choice of chains having the same number of –OH groups directly attached, the name is based on the most unsaturated chain, etc. (see section B, yellow pages).

26.4 ASSEMBLIES OF IDENTICAL UNITS

This section applies when the repeated acyclic PFS†s can be collected†. The central polyvalent radical may be cyclic, acyclic or a composite of both (see App. C). Where it is a chain containing heteroatoms, the considerations of Chapter 4 (q.v.) apply.

---

†See section 1.5. The radical may bear substituents but, for the assembly method of this section to be applicable, it must be capable of being named as a di- or polyvalent radical according to the principles of Appendix C (q.v.); see also footnote to 21.9.3.2.

Example (1)

$$HS-\underset{\underset{CH_3}{|}}{CH}-O \text{— benzene-1,3,5 ring —} O-\underset{\underset{CH_3}{|}}{CH}-SH$$

$$HC_3-CH-SH$$

2,2′,2″-(Benzene-1,3,5-triyltrioxy)triethanethiol

Example (2)

$$CH_3CH_2-\underset{\underset{O}{\|}}{C}-CH_2-O-[CH_2CH_2O]_3-CH_2-\underset{\underset{O}{\|}}{C}-CH_2CH_3$$

1,1′-[Ethylenebis(oxyethyleneoxy)]dibutan-2-one

Example (3)

$$CH_3CH_2-\underset{\underset{O}{\|}}{C}-CH_2-O-[CH_2CH_2O]_5-CH_2-\underset{\underset{O}{\|}}{C}-CH_2CH_2Cl$$

4-Chloro-1,1′-(3,6,9,12-tetraoxatetradecane-1,14-diyldioxy)dibutan-2-one

26.4.1 Basic unsubstituted structures

When one of the PFSs (see Appendix D-1.1 and section 26.0) covered by section 26.1, or sections 22 and 25 are duplicated on a symmetrical divalent base-structure, it is named by citing in order (a) the locants for the points of attachment of the duplicated PFS, (b) the name of the divalent central radical (see Appendix C) (inside parentheses when compounded of two or more individual divalent radicals strung together or substituted), (c) di, tri, or tetra, etc., as appropriate, and (d) the name of the duplicated PFS (inside parentheses if need be, to avoid ambiguity).

Example (1)

$$O\underset{\diagdown}{\diagup}\begin{matrix}CH_2CH_2CN\\CH_2CH_2CN\end{matrix}$$

3,3′-Oxydipropiononitrile

Example (2)

$$CH_2OH$$

HO—C—S—S—CH—OH
HO—CH₂   CH—OH
           CH₂OH

1,2′-Dithiodiglycerol

Example (3)

CHO   CHO
HC—⬡—O—⬡—CH
CH₂CH₂CHO   CH₂CH₂CHO

2,2′-(Oxydi-*p*-phenylene)diglutaraldehyde

Example (4)

O   O   O
CH₃C—CH₂—O   OCH₂C—CH₂C—CH₃
                C=C
CH₃C—CH₂—O   OCH₂C—CH₂C—CH₃
O   O   O

1,1′-[2,2-Bis(acetonyloxy)vinylidenedioxy]di(pentane-2,4-dione)

26.4.2 Substituted derivatives

See also Appendix C-4.3.2 and D-3 (last para).

The unsubstituted structure is named as in section 26.4.1 and that is preceded by the radical prefixes for attached groups cited in alphabetical order. Priority for numbering of the repeated unit is given to the PG and then to the site of its attachment to the central radical if there is any choice. (For example, 1,2′ is 'lower' than 1′,2).

The repeated unit with the greater number of substituents has 'unprimed' locants after the above conditions have been satisfied.

Example (1)

COOH   CH₂Br
CH₂CH₂CH—S—CH₂CH₂CH₂—S—C—I
Br   CH₂—COOH

4,4′-Dibromo-3-iodo-3,4′-(trimethylenedithio)dibutyric acid

Example (2)

$$
\begin{array}{ccc}
\overset{5}{\text{HOOC}}-\overset{4}{\text{CH}}_2 & & \text{CH}_2-\text{COOH} \\
& \overset{3}{\text{HC}}-\text{O}-\overset{3'}{\text{C}}-\text{Cl} & \\
\overset{1}{\text{HOOC}}-\overset{2}{\text{CCl}}_2 & & \text{CH}_2-\text{COOH}
\end{array}
$$

2,2,3′-Trichloro-3,3′-oxydiglutaric acid

For substituted di- or polyvalent radicals, numbering begins with 1 for the carbon atom at the outermost end of the chain (i.e. that nearest to the PFS).

Example (3)

$$
\begin{array}{ccccc}
 & \overset{\text{Br}}{|} & & \overset{\text{Cl}}{|} & \overset{\text{CH}_3}{|} \\
\text{NC}-\text{CH}_2-\text{CH}_2-\text{O}-\text{CH}_2-\text{CH}-\text{CH}_2-\text{S}-\text{CH}_2\text{CH}_2\text{CH}-\text{O}-\text{CH}-\text{CN}
\end{array}
$$

2,3′-{2′-Bromo-1-chloro[thiobis(trimethyleneoxy)]}dipropiononitrile

2,3′- is 'lower' than 2′3- and that decides which arm of the central radical has the unprimed locants. Note that the entire central radical is inside enclosing marks.
Note. In the case of duplicated esters, this assembly-style is applicable only if the diradical is linked onto the acid side of the ester. If not, then the polyvalent radical has to be named which reaches as far as the ester-group and the anions collected by di, tri, tetra, etc., following a space.

Example (4)

$$
\begin{array}{ccccccc}
 & & \overset{\text{O}}{\|} & & & \overset{\text{O}}{\|} & \\
\text{CH}_3\text{CH}_2\text{CH}_2-\text{O}-\text{C}-\text{CH}_2-\text{O}-\text{O}-\text{CH}_2-\text{C}-\text{O}-\text{CH}_2\text{CH}_2\text{CH}_3
\end{array}
$$

Dipropyl dioxydiacetate
[1,1′-omitted as obvious]

Example (5)

$$
\begin{array}{l}
\qquad\qquad\qquad\qquad \overset{\text{O}}{\|} \\
\text{CH}_2\text{CH}_2\text{CH}_2-\text{O}-\text{C}-\text{CH}_2\text{CH}_3 \\[4pt]
\\
\text{CH}_2\text{CH}_2\text{CH}_2-\text{O}-\text{C}-\text{CH}_2\text{CH}_3 \\
\qquad\qquad\qquad\qquad \underset{\text{O}}{\|}
\end{array}
$$

o-Phenylenebistrimethylene dipropionate
[not *o*-Phenylenebis(propyl propionate)]

27 Poly-PG cyclic systems

27.0 PREAMBLE

As many PGs as possible should be collected in the name ending (see Appendix D-1.1), which follows that of the cyclic system bearing them.

Example (1)

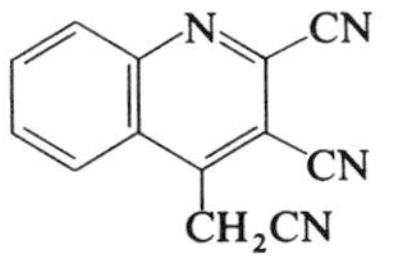

4-(Cyanomethyl)quinoline-2,3-dicarbonitrile
[not ... ylacetonitrile]

The possible applicability of section 27.6 should be explored and, when consistent with the above principle, used if it is suitable.

Example (2)

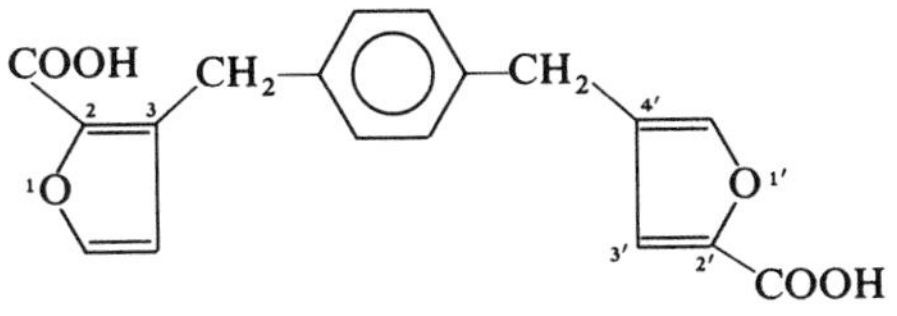

3,4'-(*p*-Phenylenedimethylene)di-2-furoic acid

However, in the event of competition for PFS (see the yellow pages), the repeated FS should be assessed for seniority on the basis of the single unit, however often repeated in the structure (cf. section 26.0, where the same principle applies to acyclic PFSs).

Example (3)

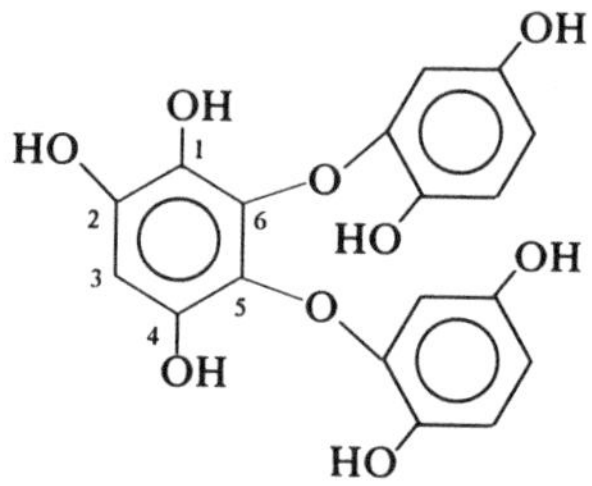

Although the two hydroquinone units can be collected in the name-ending by the assembly-method, the comparison must nevertheless be made between the two competing PFSs, viz. hydroquinone and benzene-1,2,4-triol. The latter is the senior and the name is:

5,6-bis(2,5-dihydroxyphenoxy)benzene-1,2,4-triol.

Trivial names for PFSs, e.g. those of sections 27.1.1 and 27.5.1, are preferred to more systematic synonyms, subject to any restrictive provisos added in the relevant sub-section.

In case of competition between two or more, the seniority criteria of the yellow pages (in particular section A) are applied.

When no trivial name is applicable, systematic methods are used.

27.1   CARBOXYLIC ACIDS

27.1.1   Trivially named

The following are used as a basis for naming substituted derivatives:

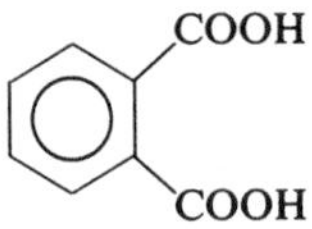

(1) Phthalic acid

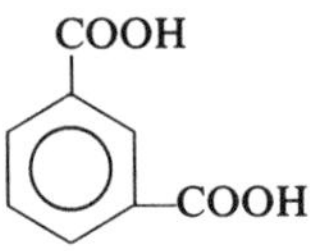

(2) Isophthalic acid

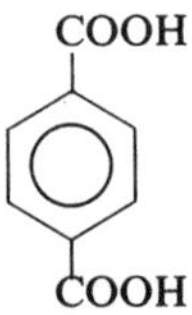

(3) Terephthalic acid

Their use as a PFS is subject to the following provisos:

(a) the ring is a benzene ring, not hydrogenated,
(b) there are no other –COOH groups directly attached to it, and
(c) there is no second benzene ring directly attached to it.

For naming in case of any non-compliance, see section 27.1.2.

27.1.2   Systematically named

Names are formed by adding the endings 'dicarboxylic acid', or 'tricarboxylic acid', as appropriate, preceded by the lowest permissible locants consistent with any fixed ring-numbering constraints, to the name of the ring or ring-system concerned (see sections 6–19).

Example (1)

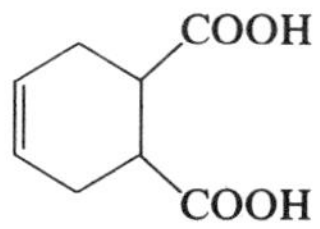

Cyclohex-4-ene-1,2-dicarboxylic acid
[not 1,2,3,6-tetrahydrophthalic acid]

Example (2)

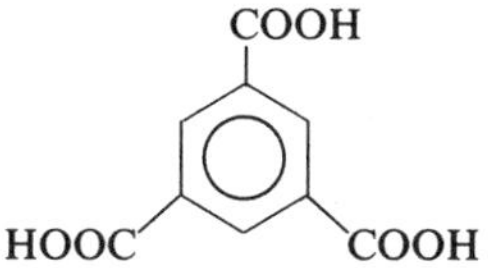

Benzene-1,3,5-tricarboxylic acid
[not 5-carboxyisophthalic acid]

Example (3)

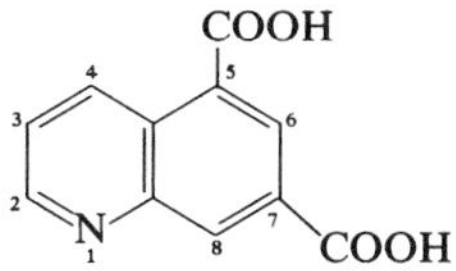

Quinoline-5,7-dicarboxylic acid

Example (4)

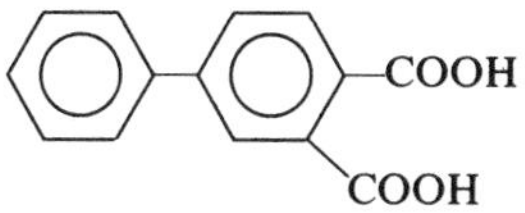

Biphenyl-3,4-dicarboxylic acid

27.2 THIOIC ACIDS

$$-COSH, \quad -CSSH$$

Names are formed by citing the name of the ring-system to which the —COSH groups are directly attached, then their locants (as low as any fixed ring-numbering constraints allow), flanked by hyphens, then 'bis(thioic acid)'. For —CSSH groups, the corresponding ending is 'bis(dithioic acid)'.

Example (1)

Benzene-1,2-bis(thioic acid)
[not 1,2-dithiophthalic acid]

27.3 FUNCTIONAL DERIVATIVES OF ACIDS

27.3.1 Non-cyclizing

The transformations from endings characteristic of the acids to those for derived nitriles, aldehydes, amides, etc., are as described in section 25 (q.v.). For the trivially named cyclic di-acids of section 27.1.1 the same stems are used in the name, the '. . . ic acid' being changed as follows:

for neutral salts and for open-chain esters, to '...ate',

for acid salts and 'half-esters' the cation name (or group name respectively) is followed after a space, by 'hydrogen' or 'dihydrogen' as the case may be, then another space and the anion name – if, after formal conversion from COOM (or COOR) to COOH yields 'collectable' COOH groups nameable according to section 27.1. If not, turn to section 24.1.0.

for acid halides to '...yl dihalide'.

For the following, each modification to the '. . . ic acid' ending implies the appropriate transformation of *both* –COOH groups:

for amides to '. . . amide',

for hydrazides to '. . . hydrazide',

for amidines to '. . . amidine',

for nitriles to '. . . nitrile', and

for aldehydes to '. . . aldehyde'.

Example (1)

Terephthalonitrile

For others, the name of the ring-system is followed by the appropriate locants (chosen as low as is consistent with any fixed ring-numbering constraints), then 'di, tri' etc., followed by the endings given in section 24 for single PGs, e.g. carboxylic acid, carbaldehyde, carbonitrile, carboxamide, etc.

Example (2)

Cyclopent-3-ene-1,2-dicarboxanilide

27.3.2 Cyclizing

27.3.2.1 Anhydrides

Those considered to be formed by internal loss of the elements of **water from individual** polybasic acids of sections 26.1.1, 26.1.2 or 27.1 are named by **replacing the word 'acid'** in the trivial name by 'anhydride'.

Examples

(1) Succinic anhydride   (2) Phthalic anhydride

(3) Cyclohex-4-ene-1,2-dicarboxylic anhydride

Other ring-systems are named with the locants for carboxylic acids followed by 'di, tri, tetra, etc., carboxylic acid' as the case may be, then, if every pair of COOH groups has been converted into an anhydride group, the COOH-locants are collected in pairs separated by a colon and the name ends with 'tetracarboxylic dianhydride' or 'dicarboxylic anhydride', as the case may be.

If however, there are also COOH groups present in the anhydride, the locants of all the COOH groups are cited, including those considered to have generated the anhydride groups, separated by commas (the set flanked by hyphens) before the ending 'tri, tetra, penta, etc., carboxylic acid'. Then, after a space, the anhydride locants and 'anhydride', or 'dianhydride', etc., as the case may be, are cited.

Example (4)

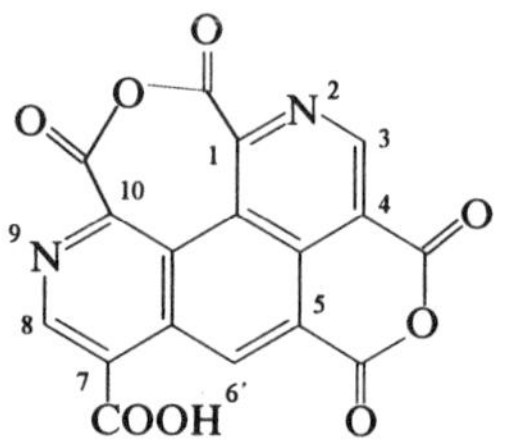

2,9-Phenanthroline-1,4,5,7,10-pentacarboxylic acid 1,10:4,5-dianhydride

27.3.2.2 *Cyclic esters*

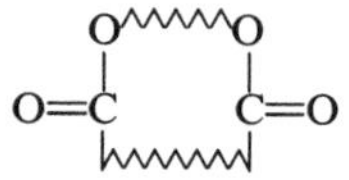

The name consists of that for a divalent radical (see Appendix C), then a space, and lastly the name of the appropriate dianion, whether trivial or systematic.

Example (1)

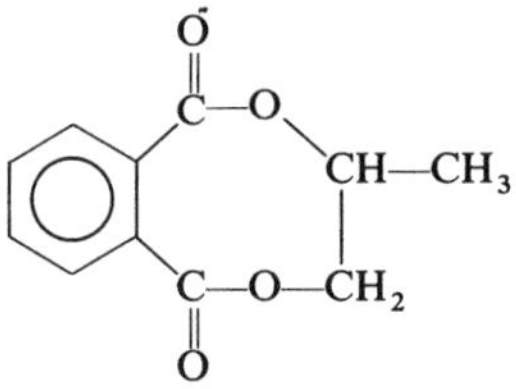

Propylene phthalate

27.3.2.3 *Imides*

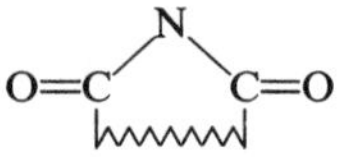

Those of the trivially named acids of section 27.1.1 are named by citing the same stem but converting the ending 'ic acid' to 'imide'.

Example (1)

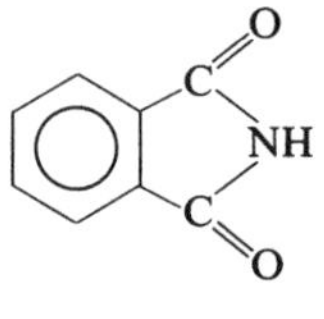

Phthalimide

For substituted derivatives the numbering of the parent acid is retained and '*N*' used for any substituents on the nitrogen atom, cited before numeral locants.

For others, the name of a parent acid of section 27.1.2 is converted from 'dicarboxylic acid' to 'dicarboximide'.

Imides derived from dioic acids of section 26.1.3 are named as heterocyclic diketones of section 27.4.3.2 (q.v.).

Example (2)

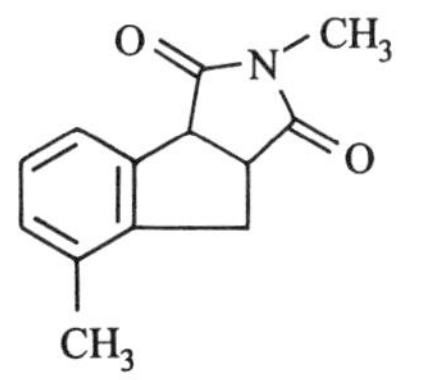

N,4-Dimethylindan-1,2-dicarboximide

Example (3)

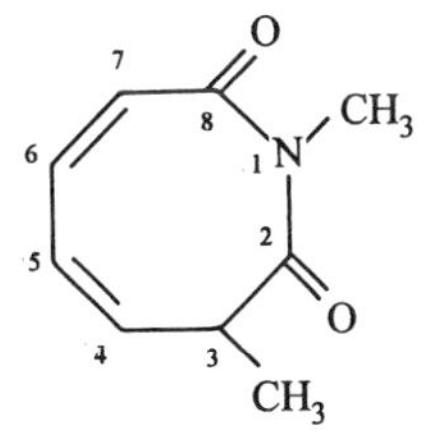

1,2,3,8-Tetrahydro-1,3-dimethylazocine-2,8-dione
[cf. ex(1) of section 26.1.3. See section 9.3 for ring name.]

27.4 POLYKETONES

27.4.1 >C=O adjacent to a ring-system

The following are cited in order: (a) the locants for attachment to the ring-system, (b) di or tri, etc., (c) the keto-chain in the form of an appropriate acyl radical (see Appendix C), and (d) the name of the ring-system.

Example (1)

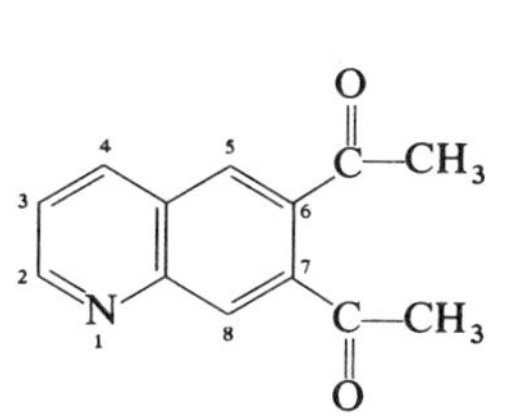

6,7-Diacetylquinoline

Example (2)

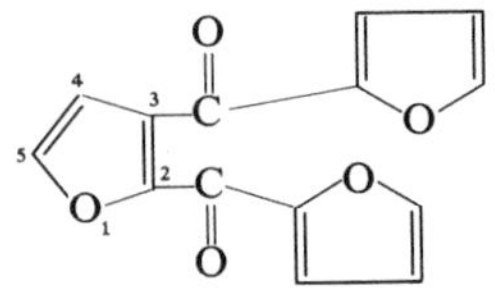

2,3-Di-2-furoylfuran

For the case of a poly-keto chain having a $>$C=O group adjacent to a ring, see section 26.2.

27.4.2 $>$C=O adjacent to two ring-systems

This structure

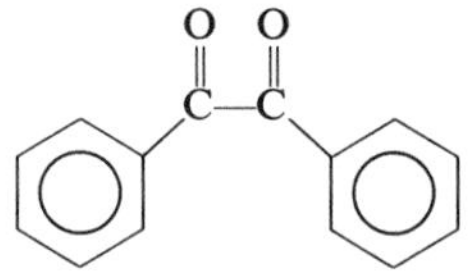

is known as benzil.
For naming substituted derivatives, one ring has its locants 'primed' and the other (bearing the same number or more substituents) does not.

Other ring systems when so duplicated (if identically substituted and connected) are named by citing in order: (a) 'di' (b) the name of the ring-system in the form of a radical prefix (see Appendix C), (c) a space, and (d) 'diketone', 'triketone', etc., according to the number of —C(=O)— groups.

Example (1)

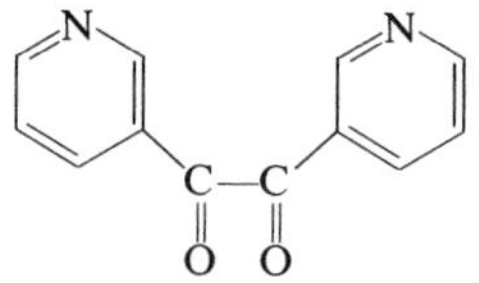

Di-3-pyridyl diketone

If the ring systems are different, or identical but differently linked or differently substituted, or both, then each is cited as a separate radical prefix in alphabetical order in the three-word radicofunctional style of section 21.3.1 (q.v.).

Example (2)

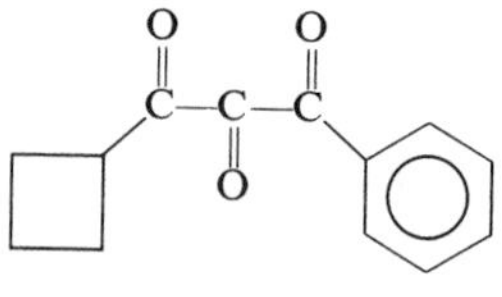

Cyclobutyl phenyl triketone

If a chain intervenes between the two keto-groups, the cyclic systems are cited as radical prefixes on the appropriately named linear dione, the chain being named on the basis of its total carbon count as in section 5.

Example (3)

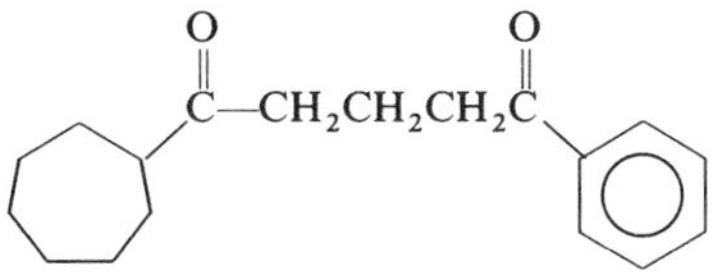

1-Cycloheptyl-5-phenylpentane-1,5-dione

27.4.3 $>$C$=$O occupying a ring position

27.4.3.1   *Quinones*

These are diketones or tetraketones derived from aromatic compounds by conversion of two or four of its ring —CH$=$ groups into $>$C$=$O with rearrangement of double bonds as shown in the examples.

Where numbering for the parent structure (see section 18.1) leaves any choice, the locant numerals for the $=$O groups are as low as fixed ring-numbering allows, taken as a set. Thereafter, any choice persisting is resolved by the pattern of substitution (see Appendix D).

Example (1)

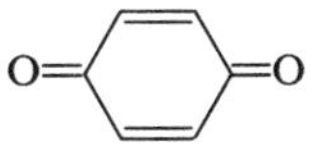

p-Benzoquinone

Example (2)

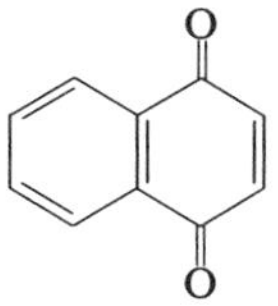

1,4-Naphthoquinone

Example (3)

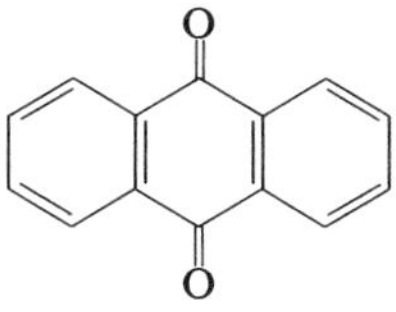

Anthraquinone

Example (4)

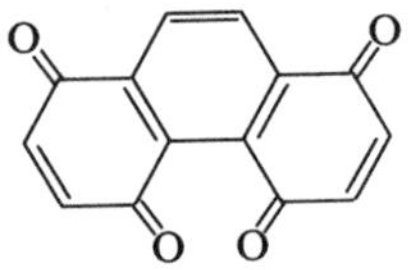

Phenanthrene-1,4:5,8-diquinone
[cf. example (4) of section 18.1]

27.4.3.2   *Non-quinonoid ring-polyketones*

Where such quinonoid structures are not present, the name is made up of that of the ring-system, the appropriate locants flanked by hyphens, and 'dione' or 'trione', etc., as the case may be. The same considerations regarding indicated hydrogen apply as for those discussed in section 16.3.

Example (1)

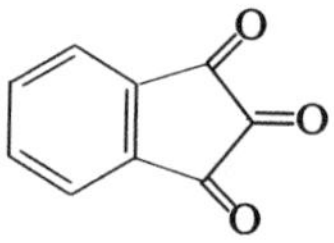

Indan-1,2,3-trione

Example (2)

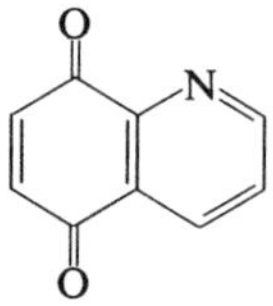

Quinoline-5,8-dione

Example (3)

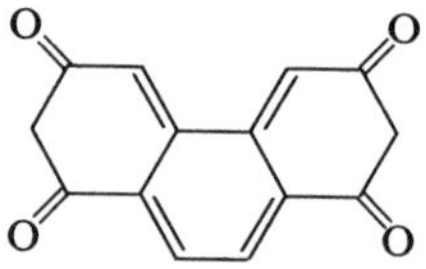

1,2,3,6,7,8-Hexahydrophenanthrene-1,3,6,8-tetrone

27.5   CYCLIC POLYOLS

27.5.1   Trivially named

The following trivial names are retained for the structures shown as a basis for naming substituted derivatives. They are used in preference to more systematically named synonyms. In case of competition amongst them, or between them and systematically named polyol PFSs, see the yellow pages, section A.

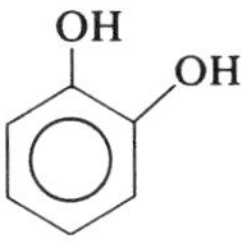

(1) Pyrocatechol

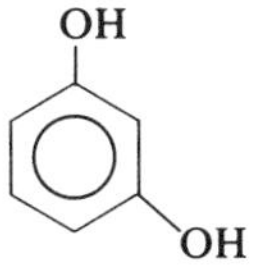

(2) Resorcinol

(3) Hydroquinone

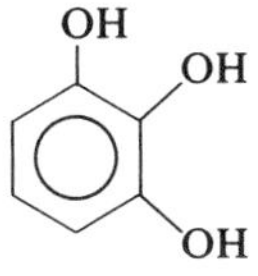

(4) Pyrogallol

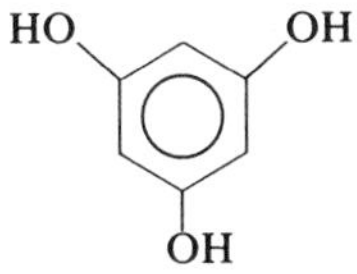

Phloroglucinol

Their use as PFS is subject to the following provisos:

(a) the ring is a benzene ring, not hydrogenated,
(b) there is no other benzene ring directly attached to it, and
(c) there are no other –OH groups directly attached to the ring.

For naming in the event of any non-compliance, see section 27.5.2.

27.5.2 Systematically named

The name is formed by citing (a) the name of the appropriate ring or ring-system, (b) the lowest locant set compatible with ring-numbering constraints for the –OH groups directly attached (flanked by hyphens), and (c) 'diol', or 'triol' etc.

Example (1)

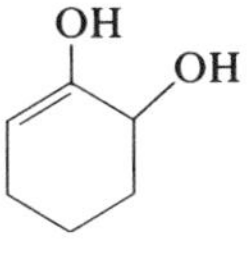

Cyclohex-1(6)-ene-1,2-diol

Example (2)

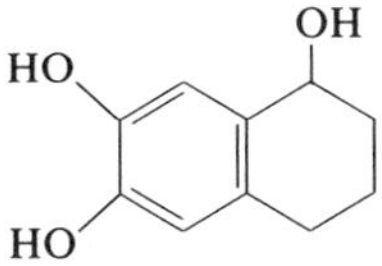

1,2,3,4-Tetrahydronaphthalene-1,6,7-triol

Example (3)

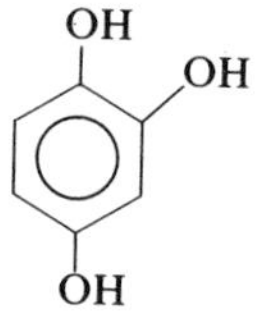

OH
OH

Benzene-1,2,4-triol

27.6 ASSEMBLIES

Where the cyclic PFS (see section 1.5 & Appendix D-1.1) is repeated by attachment to a divalent central radical‡ (whether simple or composite), the name is formed by citing the following in order:

(a) the locants on the repeated PFS, by which it is attached to the central radical, followed by a hyphen. These locants need not be the same and they can be numerals, Greek letters or italic capital atom-symbols as the case may be,

(b) the name of the central radical {if it is composite, i.e. made up of several radicals strung together (see Appendix C-4.3), or substituted, it is enclosed inside ( ) or [ ] as appropriate},

(c) 'di' or 'bis' (to avoid ambiguity), and

(d) the name of the repeated PFS.

Example (1)

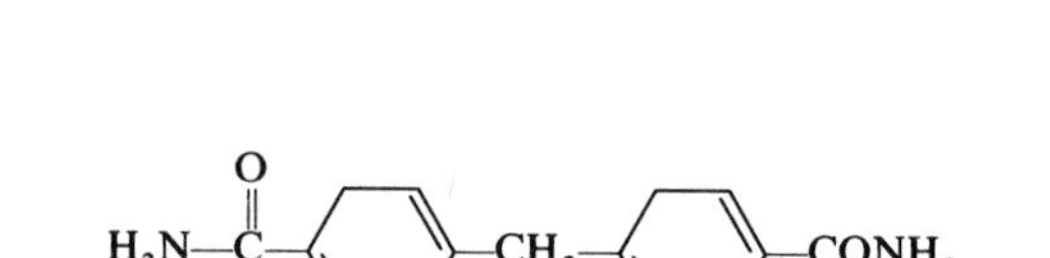

COOH
CH₃ N
N—CH₂—O—CH₂—N— —COOH
CH₃

4,6'-{Oxybis[methylene(methylimino)]}dinicotinic acid

Example (2)

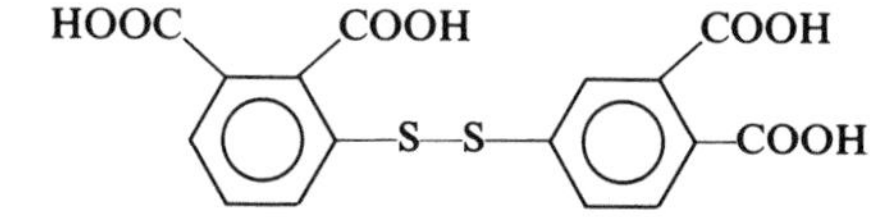

O
H₂N—C— —CH₂— —CONH₂

4,4'-Methylenedi(cyclohexa-1,4-dienecarboxamide)

Example (3)

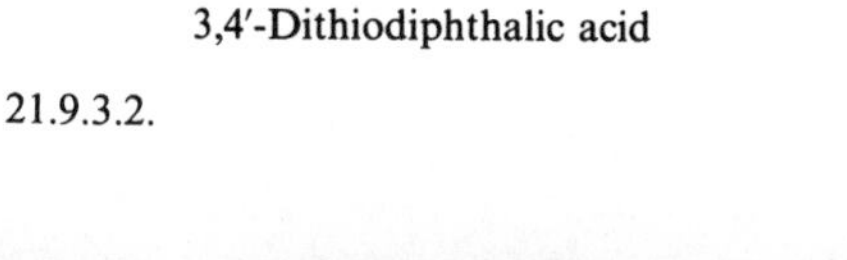

HOOC COOH COOH
—S—S— —COOH

3,4'-Dithiodiphthalic acid

‡ See the footnote to section 21.9.3.2.

See also example (2) of section 27.0.

If the repeated PFSs themselves bear substituents, radical prefixes for these groups (see Appendix C) are cited in alphabetical order, each with its appropriate locant, before the assembly name already described.

Like groups are collected wherever possible by 'di', 'tri', etc.

Example (4)

4,5,6'-Tribromo-*N*,*N'*-(cycloprop-1,2-ylene)dianthranilic acid

Example (5)

Dimethyl 3'-(3-chloroacetonyl)-2-(methylsulphamoyl)-4,4'-(vinylenedi-imino)dibenzoate

Where a cyclic system carrying a PG or PGs is directly linked to another with the same number of PGs, the seniority rules of section A, yellow pages, decide the basis of the name and the non-preferred moiety is expressed as a prefix.

If the two systems directly linked are identical (when not further substituted), the name is based on the basic assembly structure of sections 9.7, 12.2, 12.3, 13.1.1, 17.4 or 18.3.3, as appropriate. The name of the basic assembly is enclosed inside square brackets before the PG with its preceding locants and their flanking hyphens.

Lowest locants allowed by any fixed ring numbering are given to the points of attachment and, if any choice then remains, to the locants for the PGs.

The appropriate linkage prefixes are 'bi', 'ter' and 'quater' for two, three and four units of structure, respectively.

Example (6)

[4,4'-Bipyridine]-2,2'-diol

Example (7)

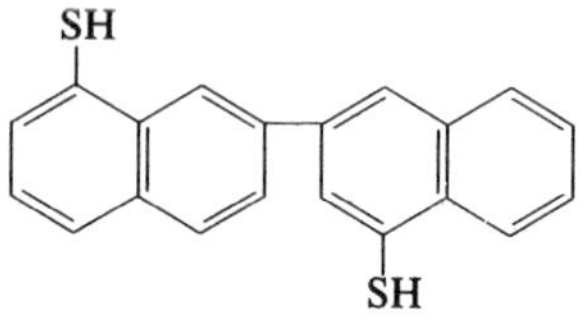

[2,2′-Binaphthalene]-4,8′-dithiol

Example (8)

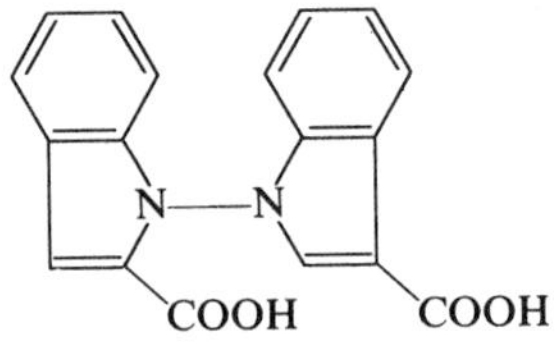

[1,1′-Biindole]-2,3′-dicarboxylic acid

If the central radical is insufficiently symmetrical,† this method cannot be used and the procedures of section 24 (or 23 if applicable), following the criteria of the yellow pages, are used instead.

Example (9)

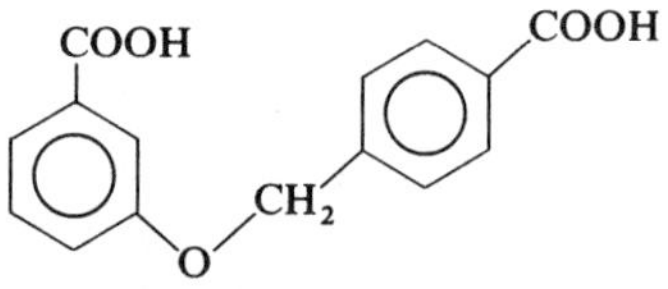

α-(4-Carboxyphenyl)-*m*-anisic acid
[Not α-carboxyphenoxy)-*p*-toluic acid]

†See the entry under 'Collected' part (iii) in section 1.5 and also the footnote to section 21.9.3.2.

28 Structures containing two nitrogen atoms directly linked

28.0 PREAMBLE

Organic molecules may contain short chains of nitrogen atoms and the naming method then depends on their length, the bond-order of their linkages, their neighbouring atoms, and whether or not they carry a charge.

Thus the grouping $-C(=O)-NH-NH_2$, whether or not the H-atoms are replaced by organic groups, is a hydrazide. This forms the name-ending if no more senior FG† is present (see section 24.13); if there is such an FG, the grouping is cited as the prefix "hydrazinocarbonyl".

Similarly the grouping $=N-NH_2$ characterizes the hydrazones of section 21.4 unless a senior FG is present, when it is cited as the prefix "hydrazono".

For N-atom chains of 3 or more the name from section 28.4 constitutes the name-ending when no senior FG's are present. When they are, the N-chain name is cited in prefix-form after resumption of the **flow-diagram** exploration at box (iii).

Chains of only two N-atoms are named by special procedures and a variety of possible combinations with other atoms is possible. Only those commonly met are considered here. For naming the rarer species, consult CNAS.

28.1 HYDRAZINES

$$R_1 \diagdown \qquad \diagup R_3$$
$$N-N$$
$$R_2 \diagup \qquad \diagdown R_4$$

R^{1-4} may be the same or different; each can also be H.

Hydrazines are treated as senior to ethers in Table 1 but not to imines. If an imine or more senior FG is present, the structure is represented to the **flow-diagram** at Box (iii) and the "YES" route from Box (iv) explored. The group is then expressed in the name as "hydrazino", the N-attachment atom being numbered 1; the other 2.

†FG defined in section 1.5.

Example (1)

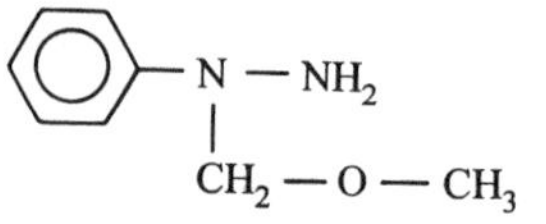

1-Methoxymethyl-1-phenylhydrazine

Example (2)

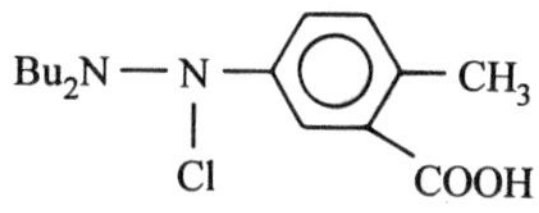

5-(2,2-Dibutyl-1-chlorohydrazino)-*o*-toluic acid

28.2 AZO-COMPOUNDS R^1–N=N–R^2

(i)   If neither R^1 nor R^2 carry FGs and R^1 and R^2 are identical when their attached groups are removed, the name is based on that of the structure RH preceded by 'azo'. Locants on the two groups are distinguished by priming, the unprimed side being chosen by Yellow pages criteria in so far as they can be taken to apply.

Example (1)

$$ClCH_2CH_2 —N = N — CHCl — CH_3$$

1,2'-Dichloroazoethane

Example (2)

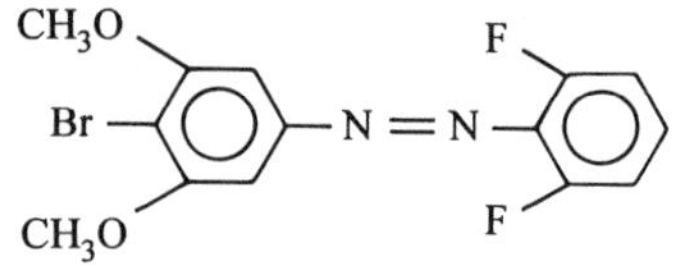

4-Bromo-2'-6'-difluoro-3,5-dimethoxyazobenzene

(ii)   If R^1 and R^2 carry the PG in the same numbers and are identical after removal of all attached groups, the naming method is that of section 26.4 for acyclic R or section 27.6 for cyclic R, the central divalent radical —N=N— being cited as 'azo'.

Example (3)

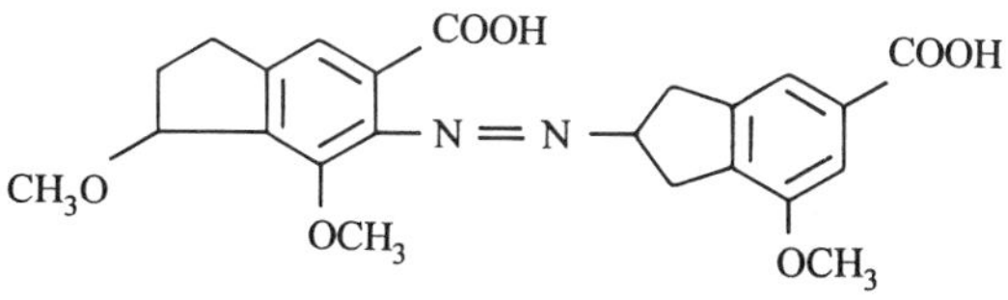

1′,7,7′-Trimethoxy-2,6′-azodi(indan-5-carboxylic acid)

(iii) If R^1 and R^2 are different after removal of their attached groups, the more senior is identified by applying the Yellow pages criteria. The senior side contains or constitutes the PFS* and the other side is cited as a radical prefix (see Appendix C) before "azo".

Example (4)

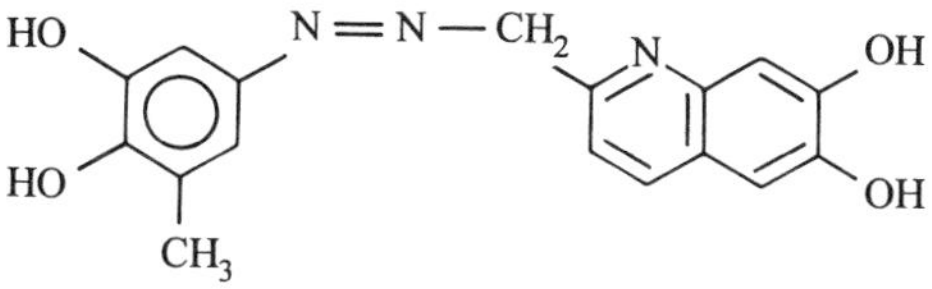

2-(5,6-Dihydroxy-*m*-tolylazomethyl)quinoline-6,7-diol

28.3   DIAZO AND DIAZONIUM COMPOUNDS

Compounds having the group $=N_2$ attached to a single C-atom are named by adding "diazo" as a prefix to the name of the parent compound, whether or not it contains a PFS. Thus, it is treated like any other group in Table 2.

Example (1)

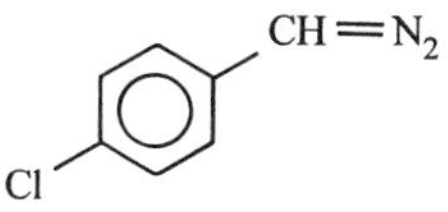

4-Chloro-α-diazotoluene

Example (2)

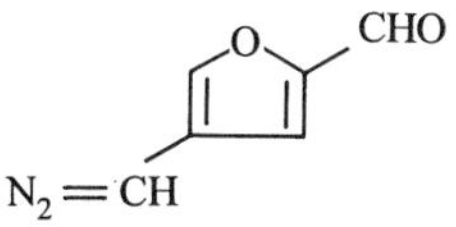

4-(Diazomethyl)-2-furaldehyde

*PFS defined in section 1.5.

The $-N^{2+}$ group (monovalent) occurs in association with a counter-anion and the name is formed by adding "diazonium" to the name of the parent structure when all attached groups have been removed. Any such groups are cited as prefixes (in alphabetical order) and the name of the anion is added after a space.

Example (3)

5-Formyl-3-furylmethanediazonium perchlorate

28.4 CHAINS OF THREE OR MORE NITROGEN ATOMS

$NH_2-NH-NH_2$ is named triazane, $NH_2-NH-NH-NH_2$ tetrazane, and so on, the Greek numerical stems of section 5.1.1 being attached as prefixes, according to the number of N-atoms, to "azane". The presence of double bonds is indicated by changing the "-azane" ending to "-azene" for one "-azadiene" for two, and so on.

If no FGs are present, any attached groups are cited in prefix-form in alphabetical order before the appropriate name for the chain of N-atoms. Numbering and citation-order follows the usual rules for C-chains (section 5.1).

Example (1)

3-(4-Bromophenyl)triazene
[lowest locant preferred for the double bond]

Example (2)

3-Butyl-1,1-dimethyltriazane
[1,1,3 preferred to 1,3,3]

When one or more FGs (imine or more senior) are present, the exploration of the **flow-diagram** is resumed at Box (iii) and the name for the N-chain is cited as the appropriate prefix formed by converting the final 'e' of the chain-name to 'o'. The point of attachment of the chain has the locant 1.

Example (3)

$$CH_3CH_2 \overset{4}{-} NH \overset{3}{-} N \overset{2}{=} N \overset{1}{-} NH \overset{}{-} \bigcirc \overset{}{-} OH$$

4-(4-Ethyl-2-tetrazeno)cyclohexanol

Appendix A

Classes of compound not covered in the book

Compounds in the following classes are named by special procedures which lie outside the scope of this book.

I   High-polymers (macromolecules).

II   Biochemicals of complicated structure, e.g. alkaloids, certain glycosides, vitamins and derivatives thereof.

Also compounds containing the following structural features.

III   . . .CO–NH . . .[CO–NH]$_x$ . . . CO–NH . . ., more complex than the oligopeptides of section 22.1, q.v.

IV

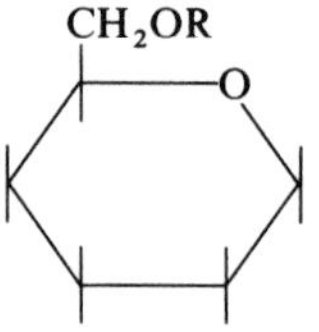

CH$_2$OR

or

CH$_2$OR

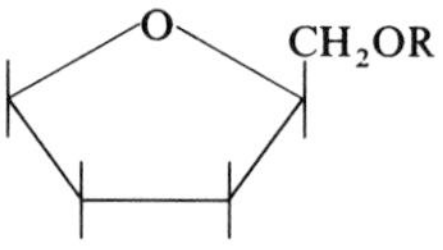

(with additional heteroatoms at more than one of the free positions).

V

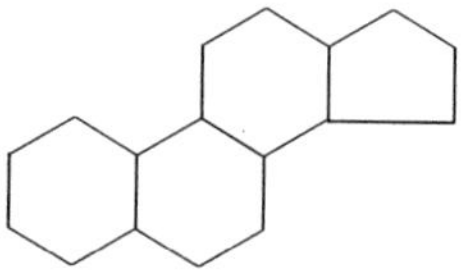

VI

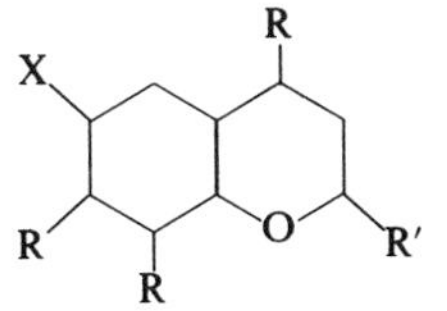

[R′ contains a C_{13} chain].

VII

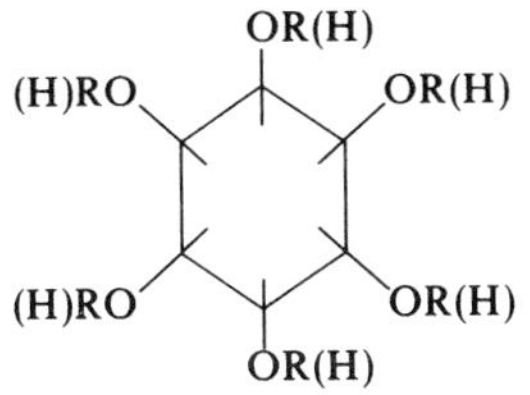

VIII

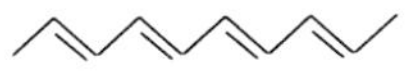

Compounds containing chains with six or more of these conjugated double bonds.

Appendix B

Rules for special-structure compounds

The compound structures of this section constitute a set of special cases, each with its own special rules. The general rules of organic nomenclature [covered by sections 1–28] are applied *only* where such special considerations allow.

B-1 CEPHALOSPORINS

Cephalosporanic acid has the structure and numbering shown:

(where X = H)

Substituted derivatives are named by citing radical prefixes (see Appendix C) in alphabetical order, each with its appropriate locant, flanked by hyphens, before the name 'cephalosporanic acid'.

Example

7-(*N*,2-Dicyclopropylacetamido)-2-(phenoxymethyl)cephalosporanic acid

Salts and esters are named by citing the name of the cation or radical, as the case may be, and then, after a space, the word 'cephalosporanate'.

If the acetoxymethyl group at position 3 is in any way modified, the compound is named on the basis of the following structure:

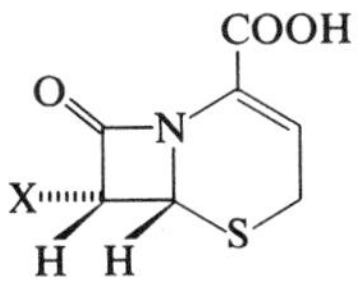

3-Cephem-4-carboxylic acid
[X=H] [same numbering]

'3-' refers to the position of the double bond. This is the position most often seen but it may be varied according to circumstance: '2-cephem' is also possible. In substituted derivatives, X is usually an amido-group.

Transformations of the –COOH to –CN, –CHO, –CONH$_2$ etc., or salt or ester-formation follow the normal conversion procedures (see for example sections 24.6–24.8, and 24.12–24.17).

Example

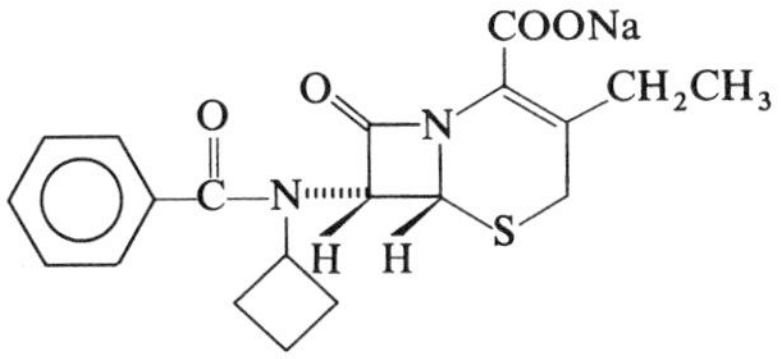

Sodium 7-(N-cyclobutylbenzamido)-3-ethyl-3-cephem-4-carboxylate

The stereochemistry shown can be assumed from the name, unless otherwise specified. Note that whether it is 7R or 7S will vary (even in the fixed configuration shown) according to the nature of individual substituents (see section 2.1).

B-2   ERGOLINE DERIVATIVES

Ergoline is the name of the structure shown with its numbering and implied stereochemistry:

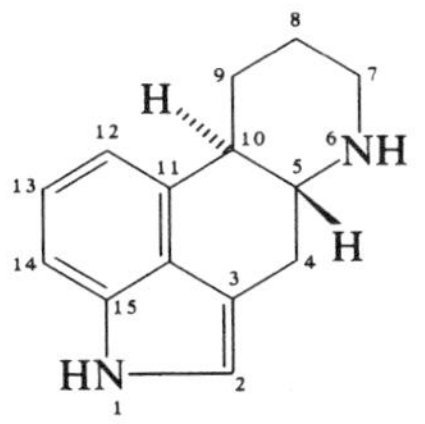

If any of the groups of Table 1 are attached directly to this structure, the PG or PGs are added to the name, taking the locant(s) from the fixed numbering shown and eliding the

'e' of ergoline before a vowel (e.g. as in the case of 'ol'). Any other attached groups are cited first in alphabetical order as radical prefixes (see Appendix C).

If any PGs are attached through an intervening group or system, the name is formed by following the route through the **flow-diagram** at the back of the book and forming the name of the PG-bearing structure according to the appropriate instructions from sections 21–28. The 'B-2' part of the molecule is then expressed as a radical prefix ergolin-*x*-yl, where *x* is the locant of the point of attachment.

Example (1)

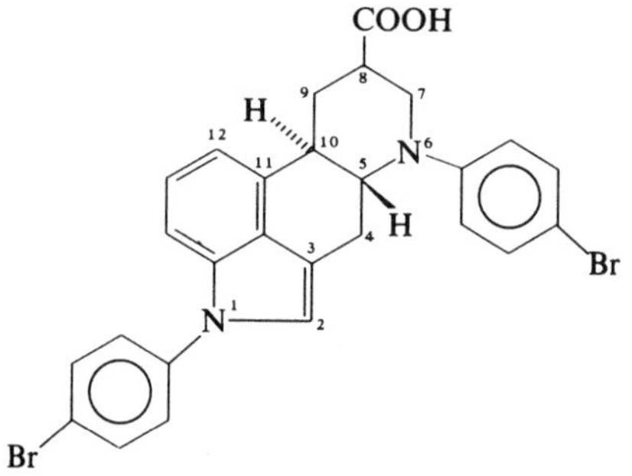

1,6-Bis(4-bromophenyl)ergoline-8-carboxylic acid

Example (2)

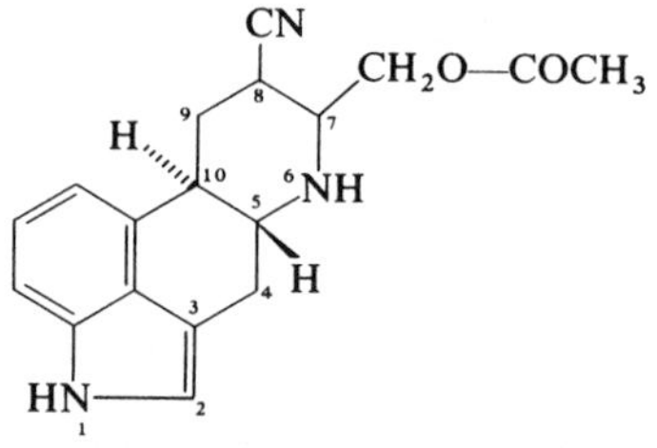

8-Cyanoergolin-7-ylmethyl acetate

Exception   If a CH_3 group is attached to position 6 *and* a $CONH_2$, $CONR_2$, or a CONHR group (where R is any radical) is attached to position 8, the name is based on D-lysergamide (same numbering).

Example (3)

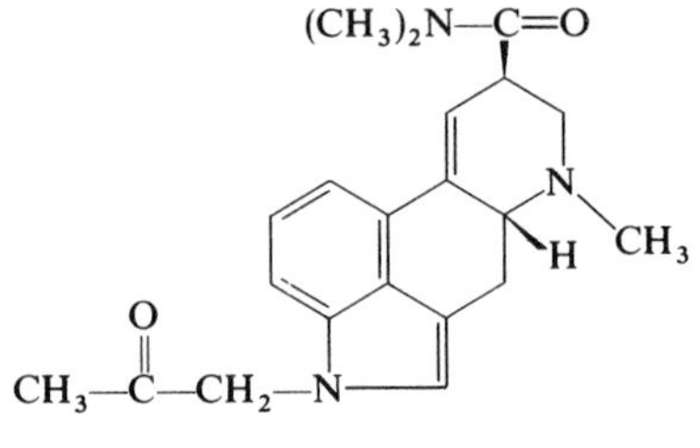

1-Acetonyl-*N*,*N*-dimethyl-D-lysergamide

If, however, a FG senior to amide is also present, the basis of the name reverts to ergoline.

B-3 MORPHINES

Derivatives in this field are named on the basis of one of the following parent structures:

(1) Morphine   (2) (−)-Morphinan   (3) (+)-Morphinan

Structure (3) is used as a basis of the name when the stereo-configuration is appropriate. Otherwise, there is a choice between names (1) and (2) according to whichever of the two structures is closest to the one under consideration. For example, loss of the -O-bridge at position 18 is a powerful incentive to use name (2).

It should be noted that morphine is a diol and, if any FG senior to alcohol is present, name (2) should be used (in radical-prefix form if applicable). In that case, N-locants are used for substituents so linked, not '17'.

Convenient usage of morphine as parent for naming may be maintained by use of the prefixes 3-deoxy, 6-deoxy or even 3,6-dideoxy when the OH groups are missing. O-locants may be used for etherified derivatives.

If the CH_3 group is not present on position 17, the corresponding $>$NH structure is termed 'normorphine', also usable as a parent and retaining morphine-numbering.

Stereochemistry is as shown unless otherwise specified.

Example (4)

(6R)-6-Deoxy-17-ethyl-7,8-dihydro-6-(methylthio)normorphine

Example (5)

(+)-N-Chloro-5,6,7,8-tetradehydro-3-hydroxy-4,5-epoxymorphinan

The ring-system can accommodate bridges across separated sites, in which case the name of the bridge (see section 19) follows the last of the radical-prefixes for attached groups (including dihydro etc.).

B-4 PENICILLINS

These are named as substituted derivatives of the parent compound:

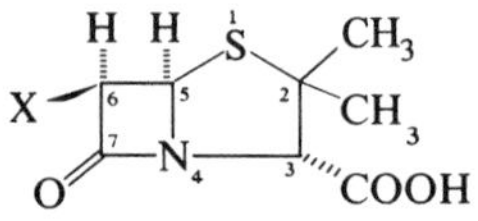

Penicillanic acid
[X=H]

Names are formed by citing in order (a) the locant for the attached radical (6) followed by a hyphen, then (b) the name of the radical prefix (see Appendix C) (enclosed inside parentheses if it begins with a numeral), and (c) 'penicillanic acid'.

Example

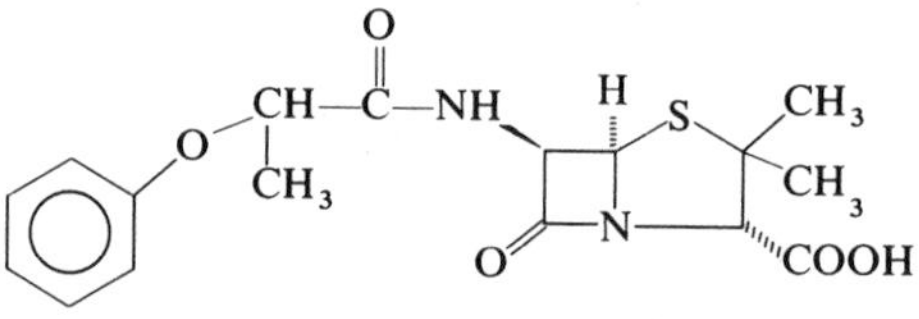

6-(2-Phenoxypropionamido)penicillanic acid

Salts and esters are named in a manner analogous to those of sections 24.7 and 24.8 (q.v.), e.g. trimethylammonium penicillanate; calcium dipenicillanate; cyclohexyl penicillanate.

Standard transformations, e.g. $-COOH \rightarrow -CONH_2$, are dealt with in the name by changing the ending 'anic acid' to 'anamide' (cf. section 24.12).

The trivial name implies the stereo-configurations shown. Any changes, as well as any new chiral centres generated by substitution, must be specified by appropriate use of *R* or *S*, preceded by their locant(s).

In the case of more drastic transformations, e.g. replacement of the S-atom by O, answer 'No' at Box (ii) and proceed through the **flow-diagram**.

B-5 TETRACYCLINES

Tetracycline itself has the following structure:

This name may be used for naming substituted derivatives, provided that the structures concerned are not too far removed from tetracycline itself. Substituents on the amide-N are characterized by N-locants whilst those on the oxygen-atoms are distinguished by use of a preceding numeral corresponding to the attached ring-locant concerned. Otherwise, substituent radical prefixes are assigned numbers as shown to indicate their position of attachment, and cited in the name in alphabetical order before 'tetracycline'.

If there is an FG senior to amide present the name is based on naphthacene (see sections 18 and 24 and, if need be, 27).

The stereochemistry is as shown unless otherwise specified in the name by means of a locant and R or S. Whether it is R or S at a particular site will depend on the substitution at that site. For example, replacement of the CH_3 at position 6 by SiH_3, with no change in the configuration shown, would change the S to R.

Example

12-*O*-Methyl-*N*-(*p*-tolyl)tetracycline

Appendix C

Radicals (attached groups)

C-0.1 NOMENCLATURE OF RADICALS

Groups having any unpaired electrons, such as triphenylmethyl $[C_6H_5]_3C\cdot$ were known hitherto as 'free radicals' but the term 'free' is nowadays tending to be dropped. Accordingly, current IUPAC recommendations are that the term 'radicals' be no longer used in discussions of molecular structure to refer to attached groups. This book, however, does not deal with what used to be called 'free radicals' and it was largely drafted before the IUPAC caveat was issued. Thus, the terms 'radical' and 'attached group' are here used interchangeably and each may refer to a monatomic or polyatomic species.

C-0.2 RECOGNITION OF RADICALS

A radical may be disguised by substitution. Thus the species CH_3- is called 'methyl' but this name may be retained even when any or all of its hydrogen atoms are replaced, e.g. CF_3- trifluoromethyl.

For easier identification of radicals when bearing substituents, hydrogen atoms are omitted to give a skeletal presentation but their presence on each unsubstituted named radical can be assumed. They are usually omitted also from ring-systems and their population may be deduced from the state of bonding present. A benzene ring is usually indicated by a circle inside a hexagon but alternating double bonds are shown in fused aromatic systems. Normal covalency of four for carbon is assumed throughout.

Where such hydrogen atom sites are indicated by a dash or line, this is distinguished from the site of the radical attachment-point by an arrow-head on the latter. Thus 'methyl' becomes $-\overset{\mid}{\underset{\mid}{C}}\rightarrow$ and ethylidene $-\overset{\mid}{\underset{\mid}{C}}-\overset{\mid}{C}=>$

Retention of H on the structure of a named radical is intended to imply that its **name** is not used when the H is replaced.

C-0.3 CITATION OF RADICALS IN A NAME

Each radical, whether simple or substituted, is cited in alphabetical order, preceded by its locant (see Appendix D-2) and a hyphen. Dihydro, tetrahydro, etc. prefixes are included under 'h'. If the same group is repeatedly attached directly to the *parent structure* (see Appendix D-1) it is collected in the name by means of di, tri, etc., or if substituted, by bis, tris, etc., preceded by a locant-set separated by commas.
[These are also used to avoid ambiguity, e.g. bis(cyclononyl) rather than dicyclononyl.]
Thus, the citation-order is: locant for the first-cited radical, hyphen, radical name; then any subsequently cited radicals are preceded by their locant flanked by hyphens. After the last cited radical comes the name of the parent structure or PFS*, as the case may be.

Example

3,6-Dichloro-4,4-bis(chloromethyl)-5-(dichloromethyl)piperidin-2-ol
'Piperidin-2-ol' is the PFS.

The prefixes name the groups attached *directly* to the PFS and the underlined letters are those considered for alphabetical ordering. Note the distinction between the two dichloro's—one is *not* directly attached to the parent structure but is on a side chain which is. Thus, the prefix letter which counts is that beginning the entire prefix name including all its cited sub-prefixes. Many illustrative examples occur in this book. See also the worked examples of Appendix D-3.

C-1 RADICALS DERIVED FROM CARBON CHAINS

C-1.1 Univalent

C-1.1.0 *Preamble*
The chain seniority criteria, yellow pages section B, are followed in so far as they can be taken to apply to radicals. The obvious difference is that the free-valence decides the position and numbering (1) of one end of the chain.

C-1.1.1 *Saturated unbranched*

$-\overset{\mid}{\underset{\mid}{C}}\rightarrow$ is methyl $-\overset{\mid}{\underset{\mid}{C}}-\overset{\mid}{\underset{\mid}{C}}\rightarrow$ is ethyl $-\overset{\mid}{\underset{\mid}{C}}-\overset{\mid}{\underset{\mid}{C}}-\overset{\mid}{\underset{\mid}{C}}\rightarrow$ is propyl

$-\overset{\mid}{\underset{\mid}{C}}-\overset{\mid}{\underset{\mid}{C}}-\overset{\mid}{\underset{\mid}{C}}-\overset{\mid}{\underset{\mid}{C}}\rightarrow$ is butyl.   Longer unbranched carbon chain radicals are named by changing the name-ending

*Defined in section 1.5.

of the hydrocarbon having the same carbon atom count (see section 5.1.1) from '-ane' to '-yl'. The carbon atom with the free valence is number 1 and numbering proceeds sequentially along the chain to its far end.

This numbering provides locants for substituents but the number (1) is not cited for the radical end as it is implicit.

C-1.1.2 *Saturated branched*

The following branched radicals have trivial names which are retained for the unsubstituted groups *only*:

Isopropyl	Isobutyl	*sec*-Butyl
tert-Butyl	Isopentyl	*tert*-Pentyl
Neopentyl	Isohexyl	

The italicized prefixes do not count for alphabetical order.

A further restriction on their use is that no carbon-chain must thereby be truncated, e.g. 2,3-dimethylheptane:

$$CH_3-CH_2-CH_2-CH_2-CH-CH \begin{matrix} CH_3 \\ \\ CH_3 \end{matrix}$$
$$\underset{CH_3}{|}$$

may not be rendered as 2-isopropylhexane.

Example (1)

1-(2-Bromoethyl)-4-chloro-3-methylbutyl
[Not 1-(2-bromoethyl)-3-(chloromethyl)butyl nor '...hex-3-yl', nor '...isopentyl'.]

Others, or any of the above when substituted, are named by general systematic methods, i.e. the free valence position is taken as terminal (and always numbered 1 but never cited) and the longest possible chain then identified and named as described in Appendix C-1.1.1 (q.v.).

Example (2)

$$-\overset{\mid}{\underset{4}{C}}-\overset{\mid}{\underset{3}{C}}-\overset{\mid}{\underset{2}{C}}-\overset{\uparrow}{\underset{1}{C}}-\overset{\mid}{C}-\overset{\mid}{C}-$$

1-ethylbutyl and not hex-3-yl nor even hexan-3-yl.

When, as in example (1), there is a choice of chains of equal length (4C), that chain bearing more substituents is preferred as the main chain.

C-1.1.3 Unsaturated

The following names are retained as the basis of naming substituted radicals:

$$>C=\overset{\mid}{C}\rightarrow \quad \text{Vinyl} \qquad\qquad >C=\overset{\mid}{C}-\overset{\mid}{C}\rightarrow \quad \text{Allyl}$$

and the following for the unsubstituted radical only:

$$H_2C=\overset{\mid}{\underset{\underset{CH_3}{\mid}}{C}}\rightarrow \quad \text{Isopropenyl}$$

Others are named in the same way as saturated radicals and with the same numbering rules, except that they contain the appropriate infixes 'ene', 'yne', 'adiene', etc., preceded by appropriate locant-numerals [see section 5.1.2]. Again, the senior chain must not be truncated by their mis-use (cf. Appendix C-1.1.2).

Example

$$\overset{4}{C}H_3-\overset{3}{C}H=\overset{2}{\underset{\underset{CH_2-[CH_2]_6-CH_3}{\mid}}{C}}-\overset{1}{C}H_2\rightarrow$$

2-Octylbut-2-enyl
[not 2-ethylidenedecyl. The unsaturated radical is preferred, even though shorter (yellow pages, (B-7)).]

C-1.2 Divalent

$\leftarrow\overset{\mid}{\underset{\mid}{C}}\rightarrow$ and $>C\Rightarrow$ are both called 'methylene'.

The context makes clear which is involved.

$\leftarrow\overset{\mid}{\underset{\mid}{C}}-\overset{\mid}{\underset{\mid}{C}}\rightarrow$ is called 'ethylene' (which is why ethene is preferred for the compound $H_2C=CH_2$).

Longer chains are named by adding 'methylene' to the appropriate numerical prefix, e.g.

$$\leftarrow\overset{\mid}{\underset{\mid}{C}}-\overset{\mid}{\underset{\mid}{C}}-\overset{\mid}{\underset{\mid}{C}}\rightarrow \quad \text{trimethylene} \qquad\qquad \leftarrow[\overset{\mid}{\underset{\mid}{C}}]_4\rightarrow \quad \text{tetramethylene.}$$

The following are also retained and used as a basis for naming assembly-derivatives (see section 27.6):

H
|
H—C—H
| Propylene   ←C=C→   Vinylene
←C-CH₂→   [Unsubstituted
| only]
H

—C=C⇒   Vinylidene

Radicals of form R—C⇒ or R—C— are named by converting the ending '-yl' of the
radical

R—C→ to 'ylidene', e.g. —C—C—C—C—C⇒ pentylidene.

They may be used either as divalent substituents on a single site as in, e.g.

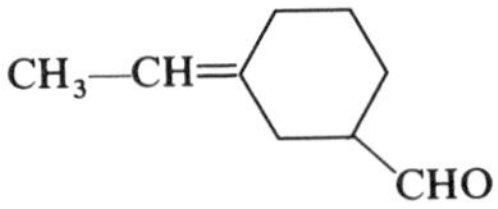

$$CH_3—CH=\bigcirc\!\!\text{CHO}$$

3-Ethylidenecyclohexanecarbaldehyde

or as the central divalent radical in an assembly, e.g.

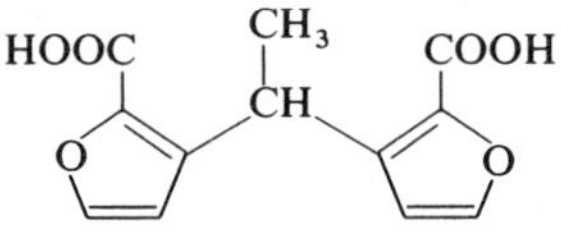

HOOC   CH₃   COOH
| CH |
(furan)   (furan)
O   O

3,3'-Ethylidenedi-2-furoic acid

C-1.3   Multivalent

Radicals of form R—C→, R—C⇒ or R—C⇒ are named by converting the end of
the name for the radical R—C→ from 'yl' to 'ylidyne', e.g.

—C—C—C—C⇒   butylidyne

Radicals having the free valencies at different carbon atoms are named by adding
'triyl', 'tetrayl' etc., as appropriate to the name of the chain on which they occur.
Note that the parent chain is terminated by a free valency at both ends. One of the two

most widely separated (or if there are only two, then one of them) is given the number 1. The other is assigned by sequential numbering of the chain-atoms. Direction of numbering is decided by (i) choosing the lower locant-set for any intermediate free-valency positions (ii) by applying criteria B-11, 12, 14, 16 of the yellow pages, in that order.

Example (1)

1-Methylhex-4-ene-1,3,6-triyl (*not* hept-2-ene-1,4,6-triyl)

Example (2)

1,1,3-Trimethylpropane-1,3-diyl

C-2   HETEROATOMIC AND COMPOSITE RADICALS

C-2.1   Monovalent heteroatomic radicals

These contain no carbon atoms. Some can be found in Table 2 and also in Table 1 columns 3 and 4. Monatomic radicals may be divided into (a) those of metals (which are usually named by changing the 'um' ending to 'o', e.g. sodio from sodium, or by changing the termination of the element name to 'io', e.g. mercurio, or by simply adding it, e.g. zincio, or by doing the same thing to a latinate form, (e.g. ferrio from iron), and (b) those of non-metals. For halogens the radical name is formed from that of the element by changing the ending '-ine' to '-o', e.g. iodo for iodine. For others it is based on that of a parent hydride in which the central element exhibits a 'normal bonding number':

$$
\begin{array}{ll}
\text{O,S,Se} & 2 \\
\text{B,N,P,As} & 3 \\
\text{Si,Ge} & 4
\end{array}
$$

The corresponding hydrides and their derived radicals are:

(1) H_2O	$HO\rightarrow$	hydroxy		(2) H_2S	$HS\rightarrow$	mercapto
(3) H_2Se	$HSe\rightarrow$	hydroseleno		(4) H_3N	$>\!N\rightarrow$	amino
(5) H_3P	$>\!P\rightarrow$	phosphino		(6) H_3As	$>\!As\rightarrow$	arsino
(7) H_4Si	$-\!Si\!\rightarrow$	silyl		(8) H_4Ge	$-\!Ge\!\rightarrow$	germyl
(9) H_3B	$>\!B\rightarrow$	boryl				

When hydrogen atoms are replaced by other groups, the first three of the above set convert to 'oxy', 'thio' and 'seleno', respectively, e.g. hexyloxy, ethylthio, phenylseleno.

Sometimes, elements display higher bonding numbers, particularly in association with oxygen, as can be seen in some of the radical-set in the following section.

C-2.2 Composite monovalent radicals

HO—S(O)$_2$→	sulpho	HO—S(O)→	sulphino
HO—S→	sulpheno	HO—(O)S(S)→	thiosulpho
$>$P→	phosphoranyl	$>$P(O)→	phosphinoyl
(HO)$_2$P(O)→	phosphono	$>$N—P(O)→	phosphonamidoyl
$>$P(S)→	phosphinothioyl	N$_3$→	azido
$>$N—N→	hydrazino	N$_2$→	diazo
O$_2$N→	nitro	ON→	nitroso
$>$N—SO$_2$→	sulphamoyl	$>$N—S→	sulphenamoyl
HO—O→	hydroperoxy	(see also Table 2)	

NB. The spelling of sulphur and its derivatives is optional in English and 'sulfur', 'sulfonic', etc. are equally correct. The 'f' spelling is the one used in N. America and it is gaining ground elsewhere as it is the form recommended for international use by the IUPAC Commission on Inorganic Nomenclature. Users may employ either form according to taste.

Composite hetero-radicals involving a carbon atom are to be found in columns 3 and 4 of Table 1. The following should also be noted:

BrC(O)→	bromoformyl	$>$N—C(O)→	carbamoyl
NCO→	cyanato	OCN→	isocyanato
—N≡CH→	formimidoyl	NC→	cyano
HC(O)N→	formamido	IC(O)→	iodoformyl
CN→	isocyano	NCS→	thiocyanato
SCN→	isothiocyanato	HC(S)→	thioformyl

Examples of substituted composite radicals are:

(CH$_3$O)$_2$P- dimethoxyphosphino; and Cl$_2$(C$_6$H$_5$)Si- dichloro(phenyl)silyl (see section 3 for silicon-chain radicals).

However, in the case of oxy radicals some contractions are used:

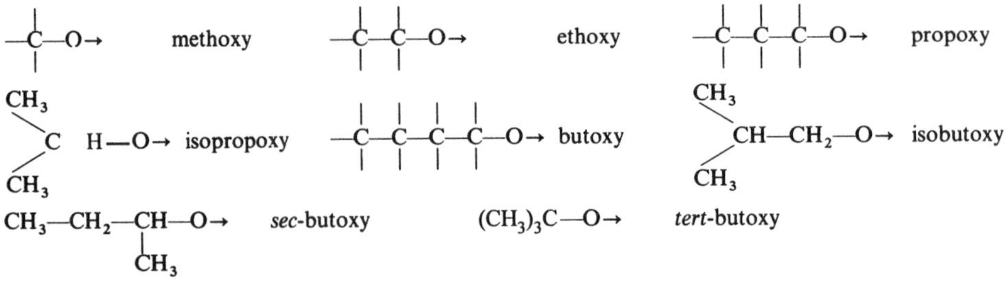

—C—O→ methoxy   —C—C—O→ ethoxy   —C—C—C—O→ propoxy

CH$_3$
  C H—O→ isopropoxy   —C—C—C—C—O→ butoxy   CH$_3$
CH$_3$ CH—CH$_2$—O→ isobutoxy
                                                                    CH$_3$

CH$_3$—CH$_2$—CH—O→ sec-butoxy   (CH$_3$)$_3$C—O→ tert-butoxy
        CH$_3$

but for others the 'alkyloxy' form is used, e.g. hexyloxy and not hexoxy. The rest, however, are used as they stand, the names for the unbranched radicals following the radical names for any substituting groups, e.g. 3,4-dichlorobutoxy.

For a more detailed treatment of -oxy radicals see Appendix C-4.3.1.

Apart from exceptional cases listed here, composite radicals are named by citing each constituent radical in turn, starting with the most remote and ending with the radical having the free valency.

Identification of the constituent radicals begins at the free valence and proceeds from it into the radical, 'biting off' as large a molecular portion as can be named by a single radical at each stage, making use of contracted forms when available.

[For a more detailed treatment see Appendix C-4.3.1.]

Example (1)

$$CH_3—O—CH_2—S→ \quad \text{methoxymethylthio}$$

Example (2)

$$CH_2{=}CH—S—S→ \quad \text{vinyldithio}$$

Example (3)

$$CH_2{=}N—\underset{\underset{S}{\|}}{C}—NH—SO_2→ \quad \begin{array}{l}\text{methylene(thiocarbamoyl)sulphamoyl}\\ \text{[not methylene(thioureylene)sulphonyl]}\end{array}$$

Of particular interest are the phosphate ester radicals:

Example (4)

$$(CH_3O)_2P(O)—O→ \quad \text{dimethoxyphosphinoyloxy}$$

Example (5)

$$(RS)_2P(O)—O→ \quad \text{(dialkylthio)phosphinoyloxy}$$

Example (6)

$$(RO)_2P(S)—O→ \quad \text{dialkoxyphosphinothioyloxy}$$

Example (7)

$$(RO)_2P(S)—S→ \quad \text{dialkoxyphosphinothioylthio}$$

etc.

C-2.3   Acyl radicals

If, by replacing the free valence by an –OH group, a trivially named acid of section 22 or 23 is generated, its stem-name is preserved in the radical name, the ending '-ic acid' being converted to '-oyl', and '-ine' to '-yl', e.g. lactoyl, arginyl.

Acyl radicals derived from trivially named fatty acids are named as follows:

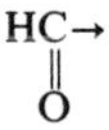

HC→   is formyl $\underset{\text{O}}{\overset{2}{-}}\overset{1}{\underset{}{C}}\to$ is acetyl $-\overset{3}{C}-\overset{}{C}-\underset{\text{O}}{\overset{1}{C}}\to$ is propionyl

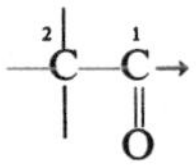

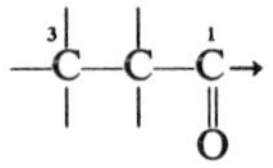

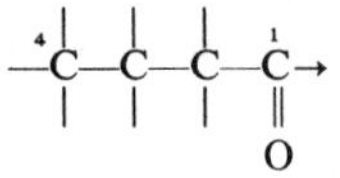

$-\overset{4}{C}-\overset{}{C}-\overset{}{C}-\underset{\text{O}}{\overset{1}{C}}\to$ is butyryl   and $-\overset{5}{C}-\overset{}{C}-\overset{}{C}-\overset{}{C}-\underset{\text{O}}{\overset{1}{C}}\to$ is valeryl

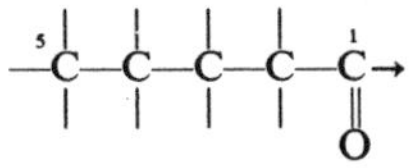

but in the case of longer chain acid radicals, whether named systematically or trivially, and also unsaturated acids, the conversion is from '-ic acid' or '-oic acid' to '-oyl', e.g.

heptanoyl for $-\overset{7}{C}-[\overset{}{C}]_5-\underset{\text{O}}{\overset{1}{C}}\to$ stearoyl for $-\overset{18}{C}-[\overset{}{C}]_{16}-\underset{\text{O}}{\overset{1}{C}}\to$

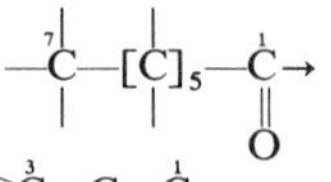

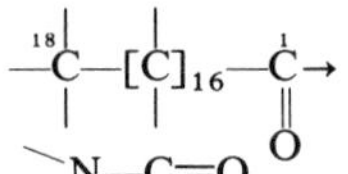

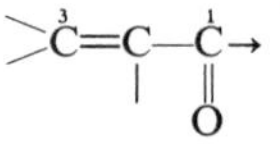

acryloyl for $\overset{3}{C}=\overset{}{C}-\underset{\text{O}}{\overset{1}{C}}\to$ oxamoyl for $\underset{\underset{\text{O}=C\to}{}}{>N-C=O}$

$-\overset{4}{C}-\overset{3}{\underset{\text{O}}{C}}-\overset{2}{C}-\overset{1}{\underset{\text{O}}{C}}\to$ is acetoacetyl, numbered as shown.

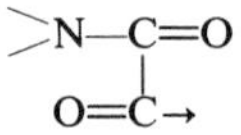

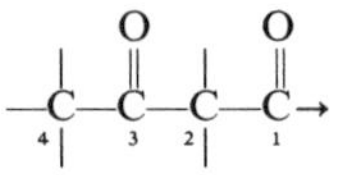

Compounds named with the ending 'carboxylic acid' generate a $-\overset{\text{O}}{\underset{}{C}}\to$ radical ending in '-ylcarbonyl'.

Example

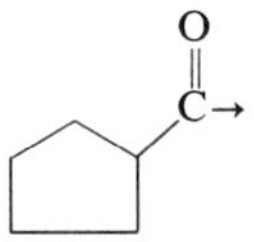

Cyclopentylcarbonyl

Radical forms deriving from functional groups are listed in column 3 of Table 1.
   The following should also be noted:

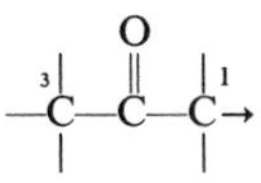

$-\overset{3}{C}-\overset{}{\underset{\text{O}}{C}}-\overset{1}{C}\to$ is acetonyl $-\overset{}{C}-\overset{}{\underset{\text{O}}{C}}-O\to$ is acetoxy.

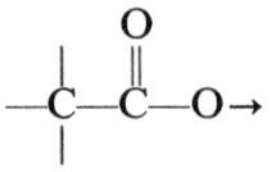

Amides having a radical-site at the nitrogen atom are named by converting the final 'e' to 'o', e.g. benzamido. However, amides deriving from amino-acids are named as acylamino-radicals, e.g.

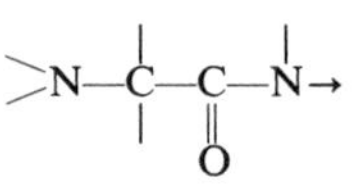

$>N-\overset{}{C}-\underset{\text{O}}{C}-N\to$ is named glycylamino.

N.B. $\underset{\overset{\text{O}}{}}{H_2N-\overset{3}{C}-\overset{2}{N}H}\overset{1}{\to}$

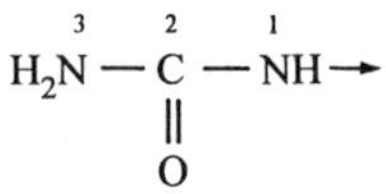

C-2.4   Ionic forms

C-2.4.1   *Cations formed by gaining a proton*

These, whether or not substituted by other groups, are named on the basis of the following parents:

	cation	prefix form
NH_4^+ (NR_4^+)	ammonium	ammonio
PH_4^+ (PR_4^+)	phosphonium	phosphonio
IH^+ (IR^+)	iodonium	iodonio
BrH^+ (BrR^+)	bromonium	bromonio

Example

$$[CH_3]_3N^+\text{—} \quad +N \underset{CH_3}{\overset{CH_3}{<}} \quad .SO_4^{2-}$$

1,1-Dimethyl-4-(trimethylammonio)piperidinium sulphate

The non-senior cation-centre is cited as a prefix on the name of the senior cationic structure.

C-2.4.2   *Anions formed by loss of a proton from an acid*

Their names are derived by changing the ending of the acid name from '-ic acid' or '-ous acid' to '-ate' and '-ite', respectively, e.g. R-carboxylic acid to R-carboxylate; sulphurous acid to sulphite.

Radical prefix forms are generated by converting the final 'e' of the anion to 'o', e.g.

$$^-O\text{—}\overset{\displaystyle O}{\overset{\|}{C}}\text{→} \quad \text{is carboxylato.}$$

It is convenient to consider the –OH group under this heading:

$$O\text{—→} \quad \text{is named oxido.}$$

Example

$$Na^+.^-O\text{—}\bigcirc\text{—}SO_2ONa$$

Disodium 4-oxidobenzenesulphonate

C-2.5   Divalent chain radicals

In addition to those mentioned in Appendices C-1.2 and C-3, the following should be noted as suitable for use, as appropriate, in symmetrical central structures for assembly names (e.g. those of section 26.4): the acids numbered (3)–(6) of section 26.1.1 convert to 'yl'; the rest to 'oyl', e.g. glutaryl for —C(=O)—[CH$_2$]$_3$—C(=O)—, but adipoyl for —C(=O)—[CH$_2$]$_4$—C(=O)—, similarly, aspartoyl.

Systematically named cases end in 'dioyl'.

Examples

(1) ←C—C→ oxalyl   (2) ←C—CH$_2$—C→ malonyl

(3) ←C—C—C—C→ tartaroyl
        | |
       OH OH

Systematically named:

←C—[CH$_2$]$_8$—C→   decanedioyl

In addition, the following should be noted:

←O→	oxy	←NH→	imino	NH‖←C→ carbonimidoyl
←O–O→	dioxy	←S→	thio	O‖←C→ carbonyl
←S—S→	dithio	←Se→	seleno	
←NH—NH→	hydrazo	O‖←Se→	seleninyl	S‖←C→ thiocarbonyl
←N=N→	azo†	Se‖←C→	selenoxo	

Note also the following examples of composite forms:

←O—CH$_2$—O→   methylenedioxy   ←O—CH$_2$CH$_2$—O→   ethylenedioxy
←O—[CH$_2$]$_3$—O→   trimethylenedioxy   ←O—C—O→   oxycarbonyldioxy‡
                                                        ‖
                                                        O

←O—S(O)$_2$—O→   sulphonyldioxo   ←NH—C—NH→   ureylene
                                                    ‖
                                                    O

C-3   CYCLIC RADICALS

C-3.1   Monovalent

C-3.1.1   General

The name of the corresponding monocyclic hydrocarbon (see sections 6 and 7.1) has its ending changed from '-ane' to '-yl', or from '-ene' to '-enyl'. The position of the radical is numbered 1, but this is implicit and so is omitted from the name.

† Note also that R–N=N→ is called 'R-ylazo', and also that C$_6$H$_5$–N=N–C$_6$H$_5$ is named azobenzene.
‡ This is usable when the repeated structure has a PG senior to –OH. Otherwise, the name is based on the ester of carbonic acid (inorganic).

Examples

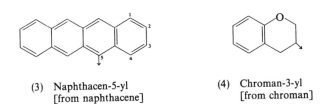

     (1) Cyclopropyl   (2) Cyclohex-2-enyl

Where the ring-system has any fixed numbering as in structures of sections 9 and 11–19 inclusive, the locant for the radical position must be included in the name, formed by adding 'yl' to that of the structure or else replacing its final 'e' by 'yl'.

Examples

     (3) Naphthacen-5-yl   (4) Chroman-3-yl
          [from naphthacene]   [from chroman]

After that, lowest locants are assigned to unsaturation and then to substituent-locants, taken as a set.

Examples

     (5) Bicyclo[2.2.2]oct-5-en-2-yl   (6) 2,2,3-Trimethylcyclopropyl
                                     [not 2,3,3-]

Exceptions:

For the following contracted forms, the radical-locant comes at the start, e.g. 1-naphthyl; not naphth-1-yl.

Examples

     (7) Naphthyl   (8) Anthryl
          [2- shown]   [9- shown]

     (9) Phenanthryl   (10) Furyl
          [4- shown]   [3- shown]

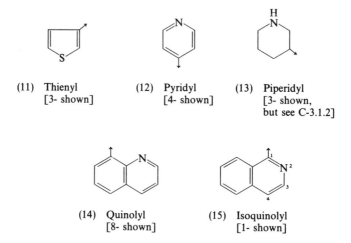

(11) Thienyl
[3- shown]

(12) Pyridyl
[4- shown]

(13) Piperidyl
[3- shown,
but see C-3.1.2]

(14) Quinolyl
[8- shown]

(15) Isoquinolyl
[1- shown]

The monovalent radical from benzene is named 'phenyl' but, when it is directly linked to another benzene ring, the radical is named 'biphenyl-x-yl', where 'x' is the locant for the free valence. For three benzene rings so linked the radical is named "terphenyl-x-yl" preceded by linkage locants (see sections 12.2 and 18.3.1).

C-3.1.2 Hydrogenated forms

These are named by means of dihydro, tetrahydro, etc., prefixes, followed by the name of the mancunide§ structure. The radical site takes precedence over unsaturation, for choice of lowest locant consistent with any fixed ring numbering constraints.

Example (1)

5,6,7,8-Tetrahydro-2-naphthyl

The following are exceptional trivial names for particular heterocyclic radicals.

Examples

(2) Piperidino
[1- only]

(3) Morpholino
[4- only]

— by contrast:

(4) morpholin-3-yl

C-3.1.3 *Ring-with-chain radicals*

The following are retained as parent radicals and numbered as shown:

Examples

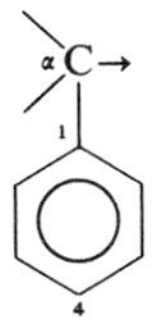

(1) Benzyl 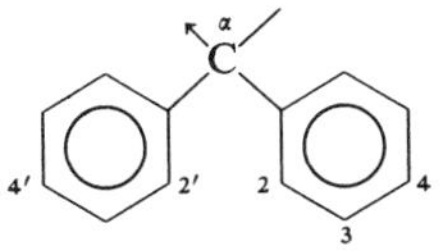   (2) Benzhydryl 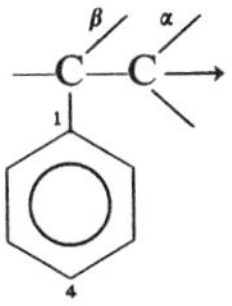   (3) Phenethyl

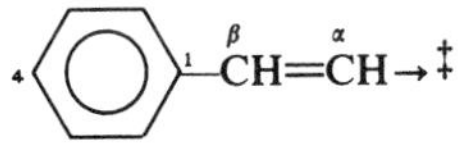

(4) Styryl† 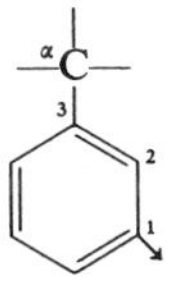   (5) Tolyl   (6) Xylyl
                         [*m*- shown]   [2,6- shown]

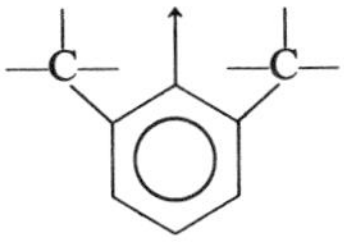

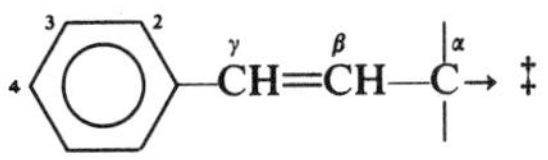

(7) Cinnamyl 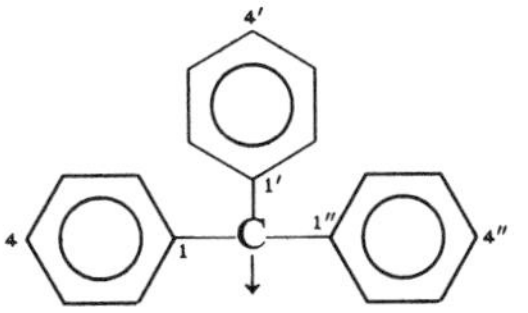   (8) Trityl

but not when their use would involve forfeiture of due seniority, such as the truncation of a chain; nor when the ring bears extra methyl groups in the case of (5) and (6).

Example (9)

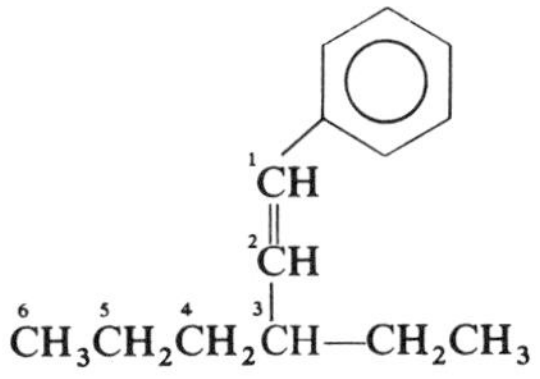

(Z)-3-Ethyl-1-phenylhex-1-ene
(not 3-styrylhexane)

Example (10)

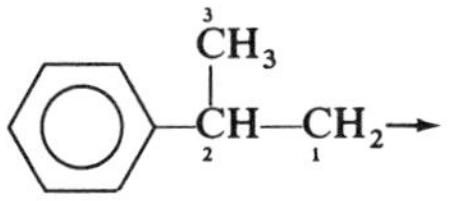

2-Phenylpropyl
(not β-methylphenethyl)

---

† Contrast the αβ-locant disposition on styrene (section 8.1).
‡ Hydrogen atoms are substitutable but may be either (Z) or (E); (see section 2.2).

NB. α-Methylphenethyl is correct, however, for C_6H_5—CH_2—$CH(CH_3)$→.
Note also that the composite radical phenyloxy contracts to phenoxy, subject to biphenylyl restriction (see also Appendix C-4.3.1).

Names for N→ radials derived from the aromatic amines of section 21.9.1 are generated by changing their final 'e' to 'o'.

Example (11)

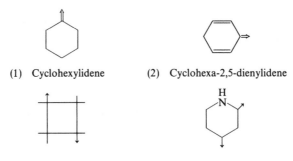

p-Toluidino

C-3.2   Divalent

When both free valencies are on the same carbon atom, the '-ane' ending of the named carbocycle is changed to 'ylidene'.

When they are on different sites, the parent structure-name is followed by the lowest locants allowed by fixed ring-numbering constraints, a hyphen, and then 'ylene'; heterocycles have 'diyl' instead.

Examples

(1)   Cyclohexylidene   (2)   Cyclohexa-2,5-dienylidene

(3)   Cyclobutan-1,3-ylene   (4)   Piperidine-2,4-diyl

Note also the contracted form: 'naphthylene'

2,3-Naphthylene shown

C-4   SUBSTITUTED RADICALS

C-4.1   Open chain radicals

This section should be read in conjunction with Appendix D.

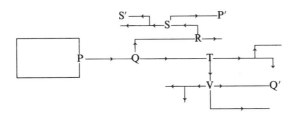

These arrows signify the direction of *exploration* and their sense is the reverse of that used in name-construction.

In the diagram, the box represents a parent structure of a complicated molecule. It consists of a senior unit, whether a chain, a ring or ring-system, bearing the PG or PGs (see definitions section 1.5). The branched line-network represents a substituted radical directly attached. Its name has to be built up by identifying the senior portion starting at the point of attachment P and exploring outwards away from the parent structure. For this purpose the atom Q attached to the parent-site P is numbered 1 and the numbering proceeds sequentially away from P. In the case of an acyclic radical, the choice of senior radical requires adaptation of the seniority criteria of section B, yellow pages, to the constraint that the free valency identifies position 1 of the radical. Thus, (B-1), (B-2) and (B-10) do not apply because PGs on a structure are expressed in prefix-form when that structure is cited as a radical prefix (see examples (1)–(5) below). However the rest do apply subject to the fixing of position 1 (cf. section 8.2 ex.7).

If the chain P–P' is assumed to be the senior, and this chain has substituents on it at Q, R and S, then the same process of identifying senior radicals when branched is carried out with outward exploration at these points. If Q–Q' is identified as the senior sub-chain attached at Q to P–P', then this Q–Q' chain undergoes similar analysis at points T and V, and so on until every last branch has been explored to its outermost tip and identified by name.

The names of sub-chains and sub-sub-chains are then compiled by stringing together the appropriate radical names formed, using Appendices C-1 to C-3. Citation at each level of substitution follows alphabetical order.

The transition from complete parent structure to various radicals are illustrated by examples (1)–(5):

Example (1)

$$\overset{6}{C}H_3CH_2CH_2-\overset{3}{C}H-CH_2\overset{1}{C}OOH$$
$$|$$
$$CH_2OH$$

3-(Hydroxymethyl)hexanoic acid

However, if it had to be named as a group attached to a unit bearing a PG senior to –COOH, its name would depend on its point of attachment.

Example (2)

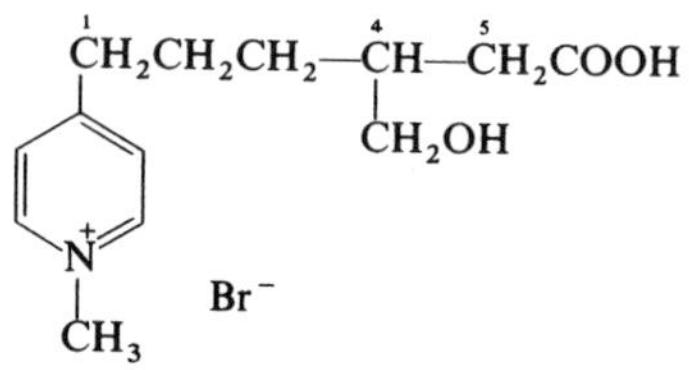

$$\overset{1}{C}H_2CH_2CH_2\overset{4}{-}\overset{}{C}H\overset{5}{-}CH_2COOH$$
$$\overset{|}{C}H_2OH$$

(4-substituted 1-methylpyridinium) Br⁻

4-[5-Carboxy-4-(hydroxymethyl)pentyl]-1-methylpyridinium bromide

By (B-15), yellow pages, carboxy is preferred to carboxymethyl in selecting the preferred pentyl radical.

Note that the group on position 4 of the pyridine, cited inside square brackets, is now named as a substituted pentyl group and that its numbering begins no longer at the –COOH group but at the attached end.

Example (3)

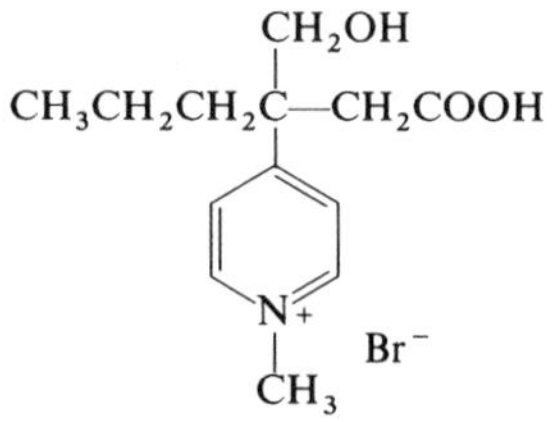

$$CH_2OH$$
$$CH_3CH_2CH_2\overset{|}{C}-CH_2COOH$$

(4-substituted 1-methylpyridinium) Br⁻

4-[1-(Carboxymethyl)-1-(hydroxymethyl)butyl]-1-methylpyridinium bromide

Example (4)

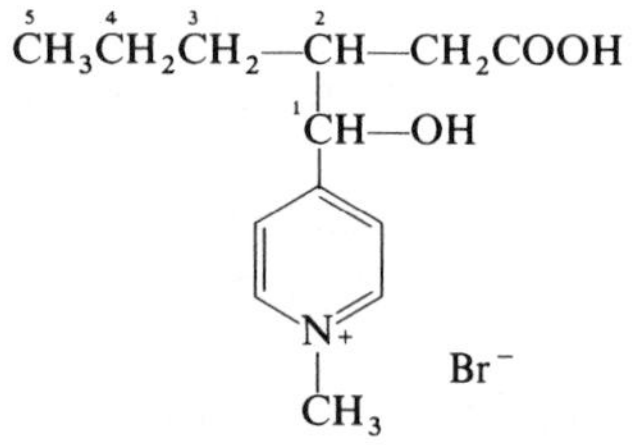

$$\overset{5}{C}H_3\overset{4}{C}H_2\overset{3}{C}H_2\overset{2}{-}\overset{}{C}H-CH_2COOH$$
$$\overset{1}{C}H-OH$$

(4-substituted 1-methylpyridinium) Br⁻

4-[2-(Carboxymethyl)-1-hydroxypentyl]-1-methylpyridinium bromide

Example (5)

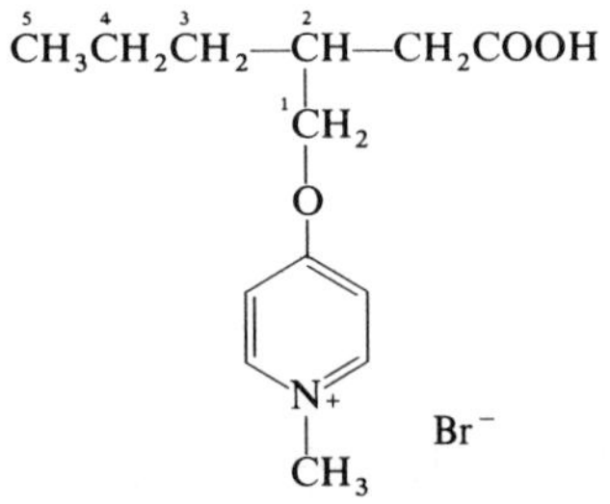

$$\overset{5}{C}H_3\overset{4}{C}H_2\overset{3}{C}H_2\overset{2}{-}\overset{}{C}H-CH_2COOH$$
$$\overset{1}{C}H_2$$
$$O$$

(4-substituted 1-methylpyridinium) Br⁻

4-[2-(Carboxymethyl)pentyloxy]-1-methylpyridinium bromide

C-4.2 Substituted cyclic radicals

Most cyclic systems have trivial names but, where systematic procedures are followed, e.g. for fused systems, then the name is generated as if the free radical position were filled by a hydrogen atom. The naming process then consists of identifying the structure bearing the free radical by applying relevant instructions from the appropriate sections with the aid of the **flow-diagram** at the back of the book.

Some information on the formation of radical names is given in the individual sections but, in general, where fixed numbering applies, the radical is distinguished by the locant for its free position before the name of the structure, with its ending converted to 'yl'.

Examples

(1) Inden-2-yl

(2) Purin-7-yl
[7*H*-purine is known simply as 'purine'.]

(3) Benzo[*a*]anthracen-3-yl

(4) Biphenyl-4-yl

If indicated hydrogen (see section 16.3) is needed to fix double bonding in a structure, this is preserved in the radical-name for formal reasons, even when the siting of the free valence might seem to render it superfluous.

Examples

(5) 1*H*-Cyclopenta[*d*]pyrimidin-1-yl

(6) 2*H*-Benzimidazol-2-ylidene

When combined in fused systems, or (in the case of fused systems of two or more rings) bridged aromatic systems, the full name-form is used, (with elision of any final 'e').

Example (7)

1,4-Dihydro-1,4-methanonaphthalen-9-yl
(not e.g. ...-9-naphthyl.)

However the bridged systems of section 19 use the same form for the radical-ending as for the open-chain hydrocarbon, e.g. bicyclo[2.2.2]oct-2-yl.

Example (8)

Bicyclo[2.2.2]oct-2-yl

C-4.3 Composite radicals

C-4.3.1 *Name-construction principles*

Things are complicated by the use of ring-with-chain trivial names, e.g. benzyl, as described in Appendix C-3.1.3. Restrictions on their use that are specified therein must be observed when identifying the senior radical in a substituted example. The principle to be followed (a) at the point of attachment to the parent structure, (b) at the point of attachment of each sub-group to the senior radical, and so on, is as follows.

Choose as senior radical the largest that can be covered by a single name.
In other words, starting at the free-valence position, bite off the senior molecular fragment that can be covered by a single name, whether trivial or systematic. Consider skeletal units basically but bear in mind the possibility of names embracing certain substituents on a main structure.

For the purpose of identifying the senior named radical, it may be helpful to add to the free valence a hydrogen atom and then re-submit the neutral structure so obtained to progress through the **flow-diagram** to examine the appropriate section for possible trivially named appropriate structures.

For naming, the endings '-yl', '-ylene', '-ylidene', '-ylidyne', and in most heteroatom cases, '-o' (e.g. anilino) characterize a radical termination as does '-io' for metal elements and cationic radicals. RO-radicals end in 'oxy' and their contracted forms, e.g. alkoxy, phenoxy, count for these purposes as a single radical. Anything ending in '-yloxy', however, counts as two: an '-yl' radical and an '-oxy' (–O–) radical.

The 'largest named bite' principle still applies basically but subject to an exceptional modification; for R—O→ radicals the procedure is as follows:

(i) Remove the oxygen atom and free valence.
(ii) Name the rest of the radical so formed, using the 'largest bite' principle.
(iii) Replace the oxygen and add 'oxy' to the radical name.
(iv) If the ending '-yloxy' can *now* be replaced by a contracted form for an unbranched chain cited in Appendix C-2.2, make that conversion.

Such contractions, however convenient and attractive, may not be used when the claims of a senior chain would thereby be overridden.

Examples

$$\overset{\uparrow}{C_6H_5{-}CH_2\overset{\uparrow}{C}HCH_2CH_3}$$

(1)   1-Benzylpropyl
             [not α-ethylphenethyl]

$$C_6H_5{-}O{-}CH_2CH_2{-}S\rightarrow$$

(2)   2-Phenoxyethylthio

$$C_6H_5{-}S{-}CH_2CH_2{-}O\rightarrow$$

(3)   2-(Phenylthio)ethoxy

$$CH_3{-}\underset{}{\bigcirc}{-}CH_2CH_2\rightarrow$$

(4)   4-Methylphenethyl

In example (4), 2-*p*-tolylethyl would be wrong because, although tolyl is senior to phenyl, this consideration has jumped the step of comparing ethyl with phenethyl. *This is the first choice to be made and phenethyl wins.*

Example (5)

$$CH_3{-}\underset{}{\bigcirc}{-}CH_2CH_2{-}O\rightarrow$$

4-Methylphenethyloxy

Here 4-*p*-tolylethoxy would not be preferred because it breaks up the contraction 'phenethyl' into 'phenyl' and 'ethyl' when following the special procedure outlined above for -oxy radicals. That would not be the best possible result at stage (ii).

C-4.3.2   Substituted composite divalent radicals for use in assembly names

The method described in sections C-4.1 and C-4.2 of forming a substituted radical name by treating the main radical as a parent structure for the purpose of considering sub-radicals, is one of general application. However, for a symmetrical array of divalent radicals strung together, the following procedure is adopted.

The most central unit is identified from Appendix C-2.4 and its name is followed by 'di' when there is only one more divalent radical on either side of it, and by 'bis' if there are more. If so, the repeated divalent radicals are cited in order as progress is made away from the centre.

Example (1)

$$\leftarrow O{-}\underset{}{\bigcirc}{-}O\rightarrow$$

p-Phenylenedioxy

Example (2)

$$\leftarrow CH_2CH_2-NH-CH_2OCH_2-\overset{\overset{\displaystyle O}{\uparrow}}{S}-CH_2-O-CH_2-NH-CH_2CH_2\rightarrow$$

[Sulphinylbis(methyleneoxymethyleneiminoethylene)]

If the central radical is symmetrically substituted, it is numbered from the free ends towards the centre. The name for the substituted portion is enclosed inside parentheses and the entire composite diradical is enclosed inside square brackets.

Example (3)

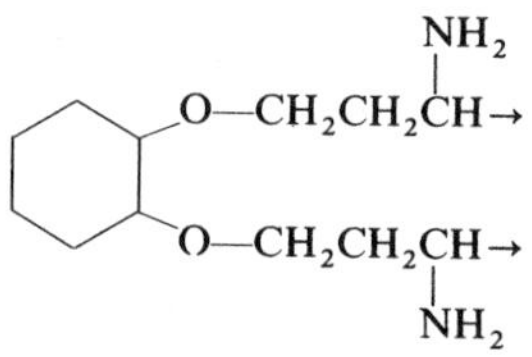

{Cyclohex-1,2-ylenebis[oxy-(1-aminotrimethylene)]}

If substitution of the central radical is not symmetrical, the assembly method cannot be used*. Instead, the PFS is chosen by following the **flow-diagram** route (xxx)→(xxviii).

*See also "Collected" section 1.5.

Appendix D   Guide to name-construction

D-1   FORMATION OF NAMES

D-1.1   Substitutive nomenclature

According to the principles of so-called 'substitutive nomenclature', organic molecules, even if complicated, can usually be regarded as consisting of:

(a) a principal unit of structure,
(b) the principal group(s) directly attached to it, and
(c) the substituent groups.

Systematic names are formed by citing the names of these components in the order (c) (a) (b). Like groups are collected by "di-", "tri-", if simple and by "bis", "tris", etc. if substituted.

Example (1)

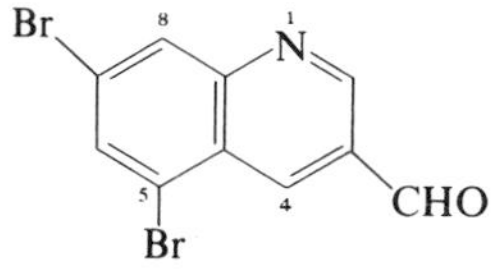

(c)   (a)   (b)
5,7-Dibromo | quinoline | -3-carbaldehyde

Example (2)

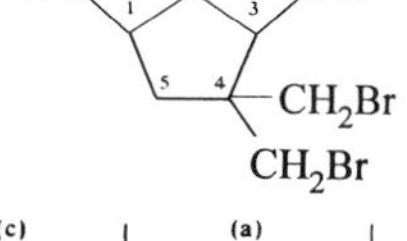

(c)   (a)   (b)
4,4-Bis(bromomethyl) | cyclopentane | -1,3-diol

Such names are completed by supplying locants for the attached groups and the PGs (as in these examples) although there are a few cases for which locants are normally omitted as being implicit [e.g. methanol, ethanol, cyclohexanol (rather than '...-1-ol'), and e.g. pentabromobenzene (1,2,3,4,5- omitted).

When the name for part (a) ends in 'e' this is elided before a vowel, e.g. propane becomes propan-2-ol.

Sometimes additional punctuation is used, such as enclosing marks, to avoid ambiguity.

Example (3)

3-(5-Chlorocyclopent-2-enyl)propoxy(hexyloxy))methanesulphonic acid

Here, the first pair of parentheses, ( ), avoids '3–5' and the second pair protects the hexyl from possible substitution by the propoxy group.

In this book the combination (a) + (b) is referred to as the **parent functional structure** (PFS).

Part (a) is chosen as appropriate from sections 3–20. If there are no FGs, the name consists simply of that for part (a) preceded by the names for any attached groups [part (c)]. Part (a) is then referred to as the **parent structure**.

In many cases however, the procedure described is complicated by the existence of contracted forms or trivial names preferred for the PFS (e.g. 2-naphthoic acid instead of naphthalene-2-carboxylic acid, for (c) alone, (e.g. morpholino instead of morpholin-4-yl), and even for (a) + (b) + (c), (e.g. salicylic acid instead of 2-hydroxybenzenecarboxylic acid).

These possibilities are taken into consideration in the **flow-diagram** but the trivial names lists in each section must be consulted before constructing names such as examples (1) and (2) above.

Such trivially named (b) + (c) combinations must be borne in mind when the seniority rules of the yellow pages are being applied. The siting of a PG (or the greatest number of PGs in cases where they proliferate) is the determining factor in naming the PFS [this is quinoline-3-carbaldehyde in the case of example (1), cyclopentane-1,3-diol in the case of example (2) and methanesulphonic acid in the case of example (3)].

Thus, in every section from 22 to 27 the naming procedure begins by identifying parts (b) and (a) in that order and then uniting them—either systematically or by an appropriate trivial name. Cases where a trivial name covers (a) (b) and (c) are limited and the progress through the **flow-diagram** should draw attention to them if applicable.

Example (4)

Here, reference to Table 1, identifies the FGs† present as amide (twice) and ether. Amide is the PG† and correct progress through the **flow-diagram** should reach section 23.1 since the amide-groups cannot be collected† in the name-ending.

Two benzamide units may be discerned but instructions in section 23.1 direct attention to section 23.1.1.5 where *p*-anisic acid is listed. Hence, by the standard transformation (see section 23.1.5.4), *p*-anisamide can be identified as the PFS. It remains only to name the attached group:

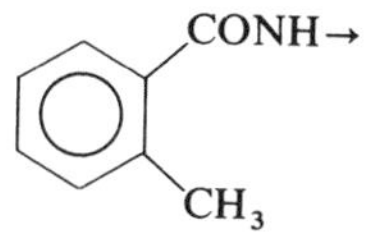

Exploring back from the position of attachment, we have successively the –NH– group (to be cited last before the 'anisamide'), the —C(=O)— and then phenyl with a methyl group attached. These give 2-(or *o*-)methylphenylcarbonylamino. Permitted combinations from Appendix C recommend *o*-toluenecarboxamido or 2-methylbenzamido. These can be further combined in *o*-toluamido.

The preferred name is thus α-(*o*-toluamido)-*p*-anisamide.

D-1.2 Radicofunctional nomenclature

PGs marked with '†' in Table 1 are covered in section 21. Their names consist of one or more radicals and a class-name, e.g. hydroperoxide, ether, sulphide, these fragments being separated by spaces.

Amines are treated under section 21 because many of their names are conveniently formed by the same processes of construction but *without* the space(s). 'Amine' has also been used as a substitutive suffix (for example, by the Chemical Abstracts Service) but that has not been the style preferred in EEC official listings and this book omits it.

Radicals may be simple (even monatomic), or complex. Their naming is described in detail in Appendix C (q.v.). When diverse, they are cited in alphabetical order.

Example (1)

$$CH_3CH_2CH_2—O—OH$$

Propyl hydroperoxide

Example (2)

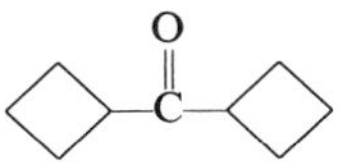

Dicyclobutyl ketone

† Defined in section 1.5.

Example (3)

3-Bromocyclohexyl 4-iodocyclopent-2-enyl sulphide

Limitations on the use of this naming style are described in section 21.

D-2 NUMBERING

D-2.1 Molecular structures

In general the senior unit of structure, whether chain or ring-system, is first identified using the yellow pages criteria, and then the numbering is considered. However, to operate criteria such as (B-11) or (B-12) the various possibilities have to be numbered and the results compared.

Many heterocyclic systems and fused carbocyclic systems have fixed numbering dictated by the requirements of orientation of the structure (see, for example, section 18.2) and/or assignment of lowest locants to heteroatoms. If such rules leave any choice, the following factors are given priority for low numbering in this order:

(a) indicated hydrogen [but Method B of section 24.18.3 can provide exceptions to this],
(b) PGs named as endings,
(c) multiple bonds in acyclic compounds and cycloalkanes,
(d) substituent groups, including hydro-prefixes,† and
(e) the substituent first cited in the name using alphabetical order.

This book contains many illustrative examples, but the following show the dependance of numbering on the various structural considerations.

Example (1)

2,4-Dibromophenol
[not 4,6-]

Example (2)

4,6-Dibromo-*o*-cresol
[the Me takes the lower number after the –OH (1 implicit)]

† Note that if you need to go this far down the list, the structure cannot be numbered until the pattern of substitution has been fully considered alongside the yellow pages criteria.

Example (3)

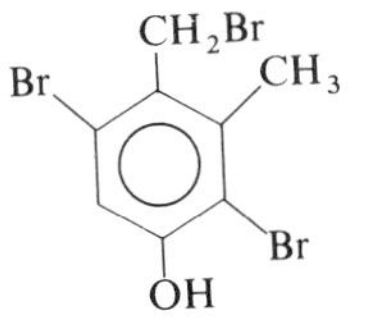

α′,2,5-Tribromo-3,4-xylenol

[The –OH takes 1 (implicit); the Me-groups take 3,4 (not 4,5) and the α for the lower-numbered Me is unprimed.]

Example (4)

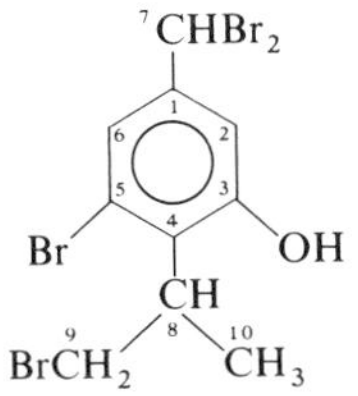

5,7,7,9-Tetrabromothymol

The numbering has to be appropriate to the style of name used.

Example (5)

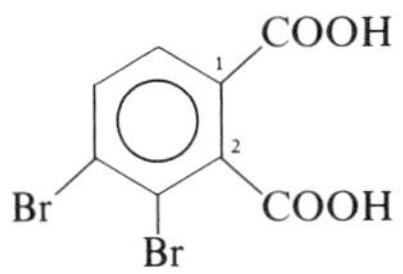

3,4-Dibromophthalic acid

[The –COOH groups assume the lowest possible locants—here 1,2. For the Br groups 3,4- is preferred to 5,6-.]

Example (6)

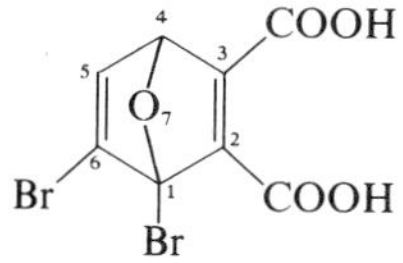

1,6-Dibromo-7-oxabicyclo[2.2.1]hepta-2,5-diene-2,3-dicarboxylic acid

Comparing example (6) with (5) we note that the aromatic ring has gone and 'phthalic acid' is no longer appropriate (just as 1,4-dihydrobenzene would not be used for cyclohexa-1,4-diene).

A von Baeyer name i.e. one in the style of section 19.1 is the correct form here and the numbering consequently begins at one or other bridgehead.

Either assigns the locant 7 to the oxygen atom and also 2,3- to the –COOH groups as well as 2,5- to the double bonds. Thus, the choice lies finally between 1,6- and 4,5- for the Br groups and 1,6- wins as being lower at the first point of difference: 1 < 4.

D-2.2 Radicals

For chains, the radical occupies position 1 of the senior chain by definition (see Appendix C-1).

For ring-systems fixed numbering is imposed with varying degrees of freedom, e.g. naphthalene: four possible sites for position 1; quinoline: only one site, viz. the nitrogen atom. Thereafter, the free-valence position takes the lowest locant possible.

Example (1)

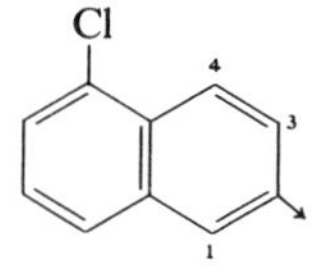

5-Chloro-2-naphthyl

Example (2)

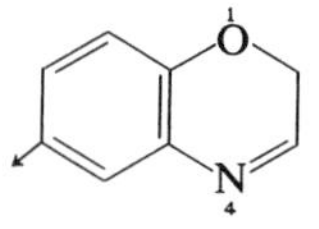

2H-1,4-Benzoxazin-6-yl

However, in single-ring hydrocarbon radicals the free valency takes the number 1 and the locants for side-chains are then made as low as possible.

Example (3)

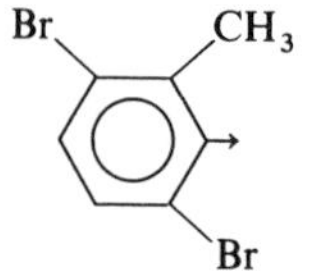

3,6-Dibromo-*o*-tolyl

Example (4)

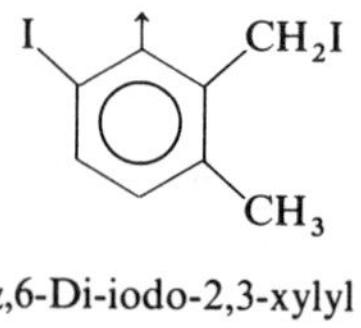

α,6-Di-iodo-2,3-xylyl

Applying what we have accrued by way of principles and guidelines from section 1 and Appendices C, D-1 and D-2, the two examples which follow provide useful illustrations of their application.

Example (1)

In example (1) the PG is –COOH and the **flow-diagram** at the back of the book takes us to section 25. The longest chain ending in –COOH is that numbered 1–5 and not the alternative path at C-2 which, although ultimately longer, is interrupted after two carbon atoms by a ring:

$$Cl_2\overset{5}{C}H-\overset{4}{C}H-\overset{3}{C}H-\overset{2}{C}-\overset{1}{C}OOH$$

(structure with NO on C-4, Br on C-3; C-2 bearing a 5-chloro-1-naphthyl group with Cl, and CH–CH₂ group with Br, attached to a pyridine ring with N, Cl, and CH₂CH₂CH₂CH₃)

On position 5 there are two chloro groups.

Although they are expressed as 'dichloro' in the name, the 'c' will count alphabetically because they are individually attached to the main chain directly.

Position 2 carries two groups, viz. a 5-chloro-1-naphthyl radical (1- preferred to 4-, 5- or 8-), and an ethyl group with a bromo group at its 1-position and a substituent at its 2-position. This, in turn, is a substituted 2-pyridyl radical.

The N of pyridine is always 1 and, for the radical position, 2 is preferred as lower than 6.

This group will appear in the name, then, as a

2-(. . . -2-pyridyl)ethyl radical.

The groups attached to the pyridine are 'chloro' on position 5 and 'butyl' on position 6. Thus the entire radical on position 2 of the valeric acid is

1-bromo-2-(6-butyl-5-chloro-2-pyridyl)ethyl
[order of citation: br, bu].

Note that the radical on the 2-position of the ethyl is enclosed inside parentheses because it starts with a numeral following a '2-'. Had the compound radical already contained parentheses, then the enclosing marks for the whole group would have been the next in the nesting order {see section 1.4(e)}, viz. [ ].

Thus, the full name is made up of the following parts:

valeric acid,
>   and the following attached groups:
>   3-bromo
>   4-nitroso
>   5,5-dichloro
>   2-(5-chloro-1-naphthyl)
>   2-[1-bromo-2-(6-butyl-5-chloro-2-pyridyl)ethyl].

When the substituent-prefixes are arranged in alphabetical order before the name of the parent acid the full name is obtained, viz.

> 3-bromo-2-[1-bromo-2-(6-butyl-5-chloro-2-pyridyl)ethyl]-5,5-dichloro-2-
> (5-chloro-1-naphthyl)-4-nitrosovaleric acid.
> ['bromo-nothing' is cited before 'bromo-something'.]

Example (2)

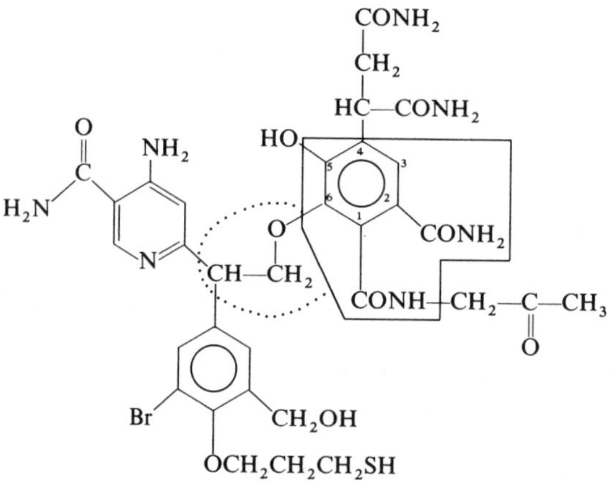

In example (2) the senior group is amide. This occurs five times in the structure but it is impossible to devise a name which will collect more than two in the same ending. The one on the left is on position 3 of a pyridine ring—a combination that can be named 'nicotinamide'. The two at the top are both joined to the same carbon chain and this unit can be named 'succinamide'. This collects two PGs and so is preferred over nicotinamide which uses only one.

The remaining two amide groups are attached to adjacent sites on the same benzene ring and this provides a combination which can be named 'phthalamide'. Once again this name subsumes two PGs and it is preferred to succinamide by the criterion C-3, yellow pages.

The full name will thus end in 'phthalamide' and this will be preceded by the names of the attached groups in alphabetical order—each preceded by its locant of attachment to the parent structure (phthalamide). For this the structure must be numbered as shown:

Example (3)

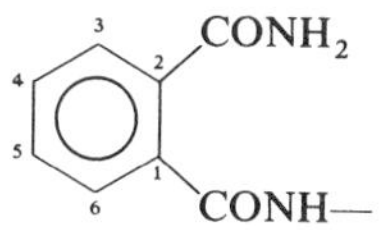

For symmetry reasons, either of the CON$\subset$ sites could be 1; the other is automatically 2.

In the case of example (2) the choice is resolved by the pattern of substitution. The locant-set $N^1,4,5,6$ is lower than $N^2,3,4,5$ at the first point of difference, so the ring is numbered as shown above.

Now, for identification of attached groups, Appendix C principles are applied, each group being considered outwards from its point of attachment.

On position 4, the succinamide residue must be considered as a substituted ethyl group, viz. 1,2-dicarbamoylethyl (propyl is not used here because its outermost carbon atom bears a $=$O).

On position 5, there is a hydroxy group.

On position 6, there is an oxy joined to ethyl.

The two groups, ethyl and oxy, are contracted to the single composite group ethoxy, but there is no such contraction for phenethyloxy (see Appendix C-4.3).

Accordingly, the first stage of outward exploration at position 6 is delineated by the dotted curved boundary, the parent structure being enclosed in an L-shaped box.

On the N^1 of the phthalamide there is a propyl chain with a $=$O on its second carbon atom. This, then, is a 2-oxopropyl group, but it is usually known as acetonyl and this is preferred. This composite radical replaces such fragmented alternatives as acetylmethyl or methylcarbonylmethyl.

The ethoxy group on position 6 is subjected to the same treatment as was the phthalamide parent for the purpose of naming attached groups. The two groups attached to position 2 of the ethoxy are:

(I) a 2-pyridyl radical [numbering principles: (a) fixed numbering constraint: N is 1, and (b) radical position is as low as possible subject to (a), so 2- not 6-].
  The pyridine ring has an amino at position 4 and a carbamoyl at position 5. The composite group is thus named 4-amino-5-carbamoyl-2-pyridyl, [citation order: a,c]. [2-Pyridyl is the preferred contracted form of pyridin-2-yl.]

(II) a benzene ring with three substituents, one of which is a substituted methyl group. Two possibilities: phenyl and *m*-tolyl. The larger is preferred. [Numbering of tolyl: the free position is 1 and the methyl-site takes 3 here (lower than 5).]
  There is then a Br on position 5 and an oxy on position 4, but this is joined to a propyl group and the contraction 'propoxy' is used for the composite group. Finally this has a mercapto group on its 3 position, numbering of the chain proceeding outwards from the point of attachment.

Thus, the group (II) on the 2-position of the ethoxy is named:

5-bromo-α-hydroxy-4-(3-mercaptopropoxy)-*m*-tolyl
[citation order: b,h,m]

The full name will be composed of the following parts:

on 4, 1,2-dicarbamoylethyl
on 5, hydroxy
on 6, ethoxy, but on its 2 position:
  (i) 4-amino-5-carbamoyl-2-pyridyl
  (ii) 5-bromo-α-hydroxy-4-(3-mercaptopropoxy)-*m*-tolyl.

These are cited in the order (i), (ii): [a before b]; so giving, on 6, 2-(4-amino-5-carbamoyl-2-pyridyl)-2-[5-bromo-α-hydroxy-4-(3-mercaptopropoxy)-*m*-tolyl]ethoxy and on N^1, acetonyl.

   The order of citation of groups attached directly to the PFS is given by ac,am,d,h and so the full name for structure (2) is:

N^1-acetonyl-6-{2-(4-amino-5-carbamoyl-2-pyridyl)-2-[5-bromo-α-hydroxy-4-
(3-mercaptopropoxy)-*m*-tolyl]ethoxy}-4-(1,2-dicarbamoylethyl)-5-
hydroxyphthalamide.

In the case of assemblies, the PFS is repeated in the structure and the same methods of name construction are used as in the above examples except that the order of citation is:

  radical prefixes for attached groups each preceded by its appropriate locant(s),
  locants for attachment of PFSs to the central polyvalent radical,
  the central polyvalent radical,
  di, tri, etc.
  the PFS.

This is set out in detail in section 26.4 (q.v.).

Table 1—Functional groups usable as name-endings

Class-name	Characteristic form as a name-ending	Radical prefix form	Molecular formula
Cation†	-ium or -onium	-io or -onio	$R^1R^2R^3R^4N^+$
Carboxylic acids	-ic acid; -oic acid; -carboxylic acid	carboxy	←COOH
Peroxycarboxylic acids	-peroxycarboxylic acid -peroxy … acid	hydroperoxycarbonyl	←C(=O)—O—OH
Carbothioic acids	-thioic acid; -carbothioic acid	thiocarboxy	←C(=O)—SH; ←C(=S)—OH
Carbodithioic acids	-dithioic acid; -carbodithioic acid	dithiocarboxy	←C(=S)—SH
Sulphonic acids	-sulphonic acid	sulpho	←S(=O)(=O)—OH
Sulphinic acids	-sulphinic acid	sulphino	←S(=O)—OH
Salts of carboxylic acids	Metal …ate; Metal …oate; Metal …carboxylate	[metal-stem (M)]io oxycarbonyl	←COOM
Salts of peroxy-carboxylic acids	Mium peroxycarboxylate; Mium peroxy …-ate;	Miodioxycarbonyl	←C(=O)—OOM
Salts of carbothioic acids	Mium O-carbothioate	Miooxy(thiocarbonyl)	←C(=S)—OM
	Mium S-carbothioate	Miothio(carbonyl)	←C(=O)—SM
Salts of carbodithioic acids	Mium carbodithioate	Miothiocarbothioyl	←C(=S)—SM
Salts of sulphonic acids	Mium sulphonate	Miooxysulphonyl-(sulphonato) [ionic]	←S(=O)(=O)—OM
Salts of sulphinic acids	Mium sulphinate	Miooxysulphinyl	←S—OM
Carboxylic anhydrides	-ic anhydride; -oic anhydride; -carboxylic anhydride	–	$\cdots C-C(=O)-O-C(=O)-C\cdots$ or $C=O$ / O / $C=O$
Lactones	-olactone; -carbolactone; -olide	–	(ring structure with R, O, C=O)

(continued on next page)

Table 1 (*contd.*)

Class-name	Characteristic form as a name-ending	Radical prefix form	Molecular formula
Esters of carboxylic acids	-yl carboxylate -yl ...oate -yl ...ate	-yloxycarbonyl	←C—OR (—COOR) ‖ O
Esters of peroxy-carboxylic acids	-yl peroxycarboxylate -yl peroxy ... oate -yl peroxy ... ate	-yldioxycarbonyl	←C—O—OR ‖ O
Esters of carbothioic acids	-yl *O*-carbothioate -yl *S*-carbothioate	-yloxy(thiocarbonyl) -ylthio(carbonyl)	S O ‖ ‖ ←C—OR; ←C—SR
Esters of carbodithioic acids	-yl carbodithioate	-ylthiocarbothioyl	←C—SR ‖ S
Esters of sulphonic acids	-yl ...sulphonate	-yloxysulphonyl	O ‖ ←S—OR ‖ O
Esters of sulphinic acids	-yl ...sulphinate	-yloxysulphinyl	←S—OR ‖ O
Esters of phosphonic acids	-yl -yl ..ylphosphonate	-..yloxy(..yloxy)phosphinoyl -di(..yloxy)phosphinoyl	OR ╱ ←P ‖ ╲ O OR
Acidic esters of inorganic acids			
Carbon acids	-hydrogencarbonate	hydroxycarbonyloxy	←O—C(=O)—OH
Sulphur acids	-hydrogensulphate -hydrogensulphite	hydroxysulphonyloxy hydroxysulphinyloxy	←O—S(=O)$_2$—OH ←O—S(=O)—OH
Phosphorus acids	-dihydrogenphosphate	phosphono-	←O—P(OH)$_2$ ‖ O
	-hydrogenphosphate	hydroxy(..yloxy) phosphinoyloxy	←O—P(OH)(RO) ‖ O
	-dihydrogenphosphite -hydrogenphosphite	dihydroxyphosphinooxy hydroxy(..yloxy) phosphinooxy	←O—P(OH)$_2$ ←O—P(OH)(OR)
Acyl halides	-yl halide; -oyl halide carbonyl halide	haloformyl	←C—X where X = ‖ Cl,Br,I, or F O
Halides of peroxy carboxylic acids	-peroxycarboxylic halide -peroxy ...ic halide	halooxycarbonyl	←C—OX ‖ O
Halides of carbothioic acids	-carbothioyl halide	halo(thiocarbonyl)	←C—X ‖ S
Halides of sulphonic acids	-sulphonyl halide	halosulphonyl	O ‖ ←S—X ‖ O

(For **neutral** esters or inorganic acids, see entries following thiols)

(*continued on next page*)

Table 1 (*contd.*)

Class-name	Characteristic form as a name-ending	Radical prefix form	Molecular formula
Halides of sulphinic acids	-sulphinyl halide	halosulphinyl	$\leftarrow$S—X, with $\parallel$O below S
Substituted ureas: Thioureas Selenoureas	-urea -thiourea -selenourea	ureido thioureido selenoureido	(H)R^1—N—C—N—R^4(H), with O(S) double bonded above C, and (H)R^2, R^3(H) below the N atoms
Amides of carboxylic acids	-amide; -carboxamide	carbamoyl	$\leftarrow$C—N$<$, with $\parallel$O below C
Amides of carbothioic acids	-carbothioamide	thiocarbamoyl	$\leftarrow$C—N$<$, with $\parallel$S below C
Amides of sulphonic acids	-sulphonamide	sulphamoyl	O double bonds above and below S: $\leftarrow$S—N$<$
Hydrazides	-hydrazide: -carbohydrazide	hydrazinocarbonyl	$\leftarrow$C—N—N$<$ with O double bonded to C, R^1(H) on first N, R^2(H) and R^3(H) on second N
Imides	-imide; -dicarboximide	iminodicarbonyl	two —C groups (each $=$O) bonded to N—R(H)
Amidines of carboxylic acids	-amidine; -carboxamidine	amidino	$\leftarrow$C—N—R^2(H) with R^1(H) on N and NR^3(H) double bonded to C
Nitriles	-nitrile; -onitrile; -carbonitrile	cyano	$\leftarrow$CN
Aldehydes	-al; -aldehyde; -carbaldehyde	formyl	$\leftarrow$CHO
Ketones	-one; ketone	oxo‡	$\begin{matrix} R^1 \\ \\ R^1 \end{matrix}C=$O or $\begin{matrix} R^1 \\ \\ R^2 \end{matrix}C=$O
Hydrazones	-hydrazone	hydrazono	$\Leftarrow$N—N$<$
Thioketones	-thione; thioketone	thioxo	$\begin{matrix} R^1 \\ \\ R^1 \end{matrix}C=$S
Oximes†	oxime	hydroxyimino (alkoxyimino etc.)	$>$C$=$N—O—R(H)
Alcohols	-ol; alcohol	hydroxy	$\leftarrow$C$\leftarrow$OH

(*continued on next page*)

Table 1 (*contd.*)

Class-name	Characteristic form as a name-ending	Radical prefix form	Molecular formula
Phenols	-ol	hydroxy	←R←OH (R is aromatic)
Thiols	-thiol	mercapto	←SH
Neutral esters of inorganic acids			
Carbonates	-carbonate	oxycarbonyloxy (monovalent)	←O—C(=O)—O←
		carbonyldioxy divalent)	←O—C(=O)—O→
Thiocarbonates	-thiocarbonate	. yloxy(thiocarbonyl)oxy	←O—C(=S)—OR
		. yloxycarbonylthio	←S—C(=O)—OR
		. ylthio(carbonyl)oxy	←O—C(=O)—SR
Dithiocarbonates	-dithiocarbonate	. ylthio(thiocarbonyl)oxy	←O—C(=S)—SR
		. ylthio(carbonyl)thio	←S—C(=O)—SR
		. yloxy(thiocarbonyl)thio	←S—C(=S)—OR
Sulphates	-sulphate	. . yloxysulphonyloxy	←O—S(=O)$_2$—OR
Sulphites	-sulphite	. . yloxysulphinyloxy	←O—S(=O)—OR
Phosphates	-phosphate	di(. yloxy)phosphinoyloxy	←O—P(=O)—(OR)$_2$
Phosphites	-phosphite	di(. . yloxy)phosphinooxy	←O—P—(OR)$_2$
Hydroperoxides†	hydroperoxide	hydroperoxy	←O—OH
Hydroxylamines	-hydroxylamine	hydroxyamino	←N—OH | R(H)
Amines†	-amine	amino	←N⟨R^1(H) R^2(H) or R^1(H)—N(R^2(H))—R^3(H)

Guanidines	Go directly to section 21.9.3.4		⟩N—C—N⟨
			    ∥
			    N—

Imines	-amine	imino	⇐N—R(H)
Ethers†	ether	-yloxy	R^1—O—R^2
Sulphides†	sulphide	-ylthio	R^1—S—R^2
Sulphoxides†	sulphoxide	-ylsulphonyl	R^1—S(=O)—R^2
Sulphones†	sulphone	-ylsulphonyl	R^1—S(=O)(=O)—R^2
Peroxides†	peroxide	-yldioxy	R^1—O—O—R^2
Disulphides†	disulphide	-yldithio	R^1—S—S—R^2

(continued on next page)

Table 1 (*contd.*)

Class-name	Characteristic form as a name-ending	Radical prefix form	Molecular formula
Oxides†	–	–	→O
Dioxides†	–	–	O ↗ ↘ O
Azides†	azide	-azido	←N$_3$

† Refer to the question in box (xxvi) of the **flow-diagram** at the back of the book. Such structures are named in the radicofunctional style under section 21. For the purposes of the questions as to whether groups are 'directly attached' in boxes (xxviii) and (xxxii) of the **flow-diagram**, this is taken to mean that the FGs in Table 1 are joined by the bond marked with an arrow-head only.

‡ Ketones. The 'oxo' prefix is applied to sites considered to have thereby lost two –H groups. Names using 'oxo' are therefore named as if these two –Hs are still present, e.g. 2-oxoindan-1-carboxylic acid; 1,4-dihydro-4-oxo-2-naphthoic acid.

Table 2—Groups which cannot have suffix-status—always cited as prefixes

Group	Prefix
Br	bromo
—Cl	chloro
—ClO	chlorosyl
—ClO$_2$	chloryl
—ClO$_3$	perchloryl
—F	fluoro
—I	iodo
—IO	iodosyl
—IO$_2$	iodyl
—I(OH)$_2$	dihydroxyiodo
—IX$_2$ (X may be a halogen or a radical) e.g.	dichloroiodo
or	diethyliodo
=N$_2$	diazo
—NO	nitroso
—NO$_2$	nitro
=N(O)OH	*aci*-nitro

Table 3—Heterocyclic prefixes for replacement nomenclature (in decreasing order of priority)

Element	Symbol	Normal bonding number	Prefix
Oxygen	O	II	oxa
Sulphur	S	II	thia
Selenium	Se	II	selena
Tellurium	Te	II	tellura
Nitrogen	N	III	aza
Phosphorus	P	III	phospha
Arsenic	As	III	arsa
Antimony	Sb	III	stiba
Bismuth	Bi	III	bisma
Silicon	Si	IV	sila
Germanium	Ge	IV	germa
Tin	Sn	IV	stanna
Lead	Pb	IV	plumba
Boron	B	III	bora
Mercury	Hg	II	mercura

1 Day
Reserve